Management for Professionals

The Springer series *Management for Professionals* comprises high-level business and management books for executives. The authors are experienced business professionals and renowned professors who combine scientific background, best practice, and entrepreneurial vision to provide powerful insights into how to achieve business excellence.

More information about this series at https://link.springer.com/bookseries/10101

David Miller

Managing Agile Business Technology

The Business and Technology Relationship Model in Practice

Springer

David Miller
Derby, UK

ISSN 2192-8096 ISSN 2192-810X (electronic)
Management for Professionals
ISBN 978-3-030-90597-2 ISBN 978-3-030-90598-9 (eBook)
https://doi.org/10.1007/978-3-030-90598-9

This Springer imprint is published by the registered company Springer Nature Switzerland AG.
The registered company address is: Gewerbestrasse 11, 6330 Cham, Switzerland

Author's Note

Making business technology work in what can be a complex, ambiguous, and dynamic world is not easy and our failures might lead people to think that our ability to manage the technology hasn't kept pace with the technology itself. The technology now has many agile characteristics but our management processes are sometimes seen as an obstacle to be overcome. We must change the way we work and the book sets out why that's the case, how this can be achieved, and the many benefits. The reader is given a conceptual solution and implementation choices. It has been written for business managers, business technology managers, and technology service providers (TSPs) but in a language that hopefully everybody will understand. This is important because in an age when everyone is affected by business technology, at all levels right across the extended enterprise, we ideally need all stakeholders to contribute and collaborate in its management and governance. It's also important because, given the opportunity, good ideas can come from the most unlikely places. DM

Preface

IT service providers have never had a complete picture of how well they are doing, and some might add that they have never had an up-to-date and complete picture of what they were supposed to do. We need to change the way we work.

In the frequent absence of a detailed and complete specification of what was required, packaged solutions, methodologies and maturity models became the order of the day. In organisations where little changed, this almost worked but success rates remained stubbornly low, even where specifications existed. As globalisation took hold, as the business landscape became more complex, as businesses had to become more dynamic and distributed, as it became impossible to know who to gather together to thrash out a problem, and as business demanded the earlier delivery of working solutions, our working methods have somehow failed to keep up. We need to change the way we work.

The modern world is complex, ambiguous, and dynamic. It is characterised by turbulent markets, shorter product lifecycles, disruptive technologies, industry structure change, high staff turnover, changes in employment status, increased home working, and changes to other working practices and daily life brought about by pandemics—or the possibility thereof, the digital economy, and the emergence of types of artificial and non-conscious intelligence, etc. Most businesses now involve technology of many kinds, may have multiple technology service providers (TSPs), and customers and suppliers will have their own technology arrangements. As a consequence, our world is always presenting multiple challenges and opportunities that require a technology solution of one kind or another, and it's not unusual for something new to impact work that is in progress more than once prior to its implementation.

It can be argued that the business world has always been in a state of flux and it is surprising then that from the business perspective, there has never been a coherent business technology management and governance model that takes account of the true nature of our world. Instead, we mostly have just a bricolage of IT methods and standards for IT delivery that are as opaque to business people as is sometimes the technology itself. Perhaps for this reason, the last 50 or more years of IT management has been characterised by uncertain processes yielding unpredictable outcomes. To adopt a motoring analogy, IT management methods applied over this time have been like driving by occasionally controlling what is happening under the bonnet but without really looking where we are going except at the start and occasionally at the end of our journey. Whilst the advent of Agile IT methods may have helped us to push down on the accelerator pedal, they have offered little in the way of navigation, steering or hindsight. As technology of many kinds determines, and gets embedded within, our business processes, we need to change the way we work.

This book offers some insight into how we could do that.

Derby, UK David Miller

Introduction

The widespread dissatisfaction with existing IT management methods and the recognition that something else was needed for the emerging and future worlds were the two drivers behind my work.

What has emerged from that work is exciting and universally appealing. The book describes a breakthrough model which suggests a new management and governance philosophy and discusses the potential advantages offered by a conceptual management system based on these.

The new model, the Business and Technology Relationship Model (BTRM), is research based, peer reviewed and fully describes the relationship between business and technology. Its discovery has been described by one industry analyst as 'mind shifting' and 'beautiful' providing simple definitions for service quality, alignment, agility, and governance that have hitherto escaped adequate definition. It explains our problems with traditional methods, democratises the management and governance of business technology to maximise collaboration at all levels, and lends itself to introducing automation.

Active Management and Governance is a conceptual Big Data management system, developed from the BTRM, that illustrates the potential for real-time, real-world insight and control that previously couldn't be contemplated. This, combined with its focus on value creation as well as value delivery, will facilitate automation and continuous transformational change and will do so in the context of current, emerging, and future technologies. How it can do this is described in full with enough information to plan an implementation, understand where AI can be leveraged, and how it can hasten a 'Very Agile' world and provide a control for self-managing systems.

The benefits will be felt in all organisations. No judgement or prediction has been made in terms whether traditional IT management methods will prevail within this new management framework or to what extent but they will certainly change. In fact, Active Management and Governance may even change the way in which we manage the business itself.

Why a book rather than a product? Because of the universal appeal of the concept there may be stages of adoption ranging from a management philosophy that just changes the way we think about managing business technology, or a customisation of an existing platform of some kind, to a specialist and sophisticated software product. There may be many variations on these themes depending on the developer's industry focus and specialist insights, and some may already be appearing in the market place. Hence it is hoped that a book which describes the philosophy and important concepts in some detail rather than an interpretation based on a restricted view will have more appeal and encourage greater innovation and wider adoption.

Chapter 1 looks back at the way technology has been managed historically, the difficulties that this has created and why it was necessary to re-think the problem. Chapter 2 describes the BTRM and how it changes the management storyline. Chapter 3 introduces Active Management and Governance to describe how the BTRM might be implemented in practice. Our thinking is often constrained by what has gone before and so Chap. 4 explains how the business landscape can be analysed in ways that have not previously been possible. Chapter 5 addresses the growing importance of AI and considers various pathways of opportunity for a business wishing to exploit learning algorithms and other non-conscious intelligence to enhance Active Management and Governance. Chapter 6 considers the management philosophy of Active Management and Governance and its potential, not just for managing business technology, but as a universal business management framework which is consistent with corporate governance practices.

Contents

About the Author

David Miller, **PhD FBCS FIET FCIM CITP CEng** David Miller's business life started in retail banking but it was at a manufacturing company, as a young management trainee when he taught himself to use Fortran to write a program to plan the production of five factories, where his career in information technology was launched. He has since worked in many business sectors in the UK and internationally as a manager or management consultant. Over a period of more than 50 years, his professional career has embraced most aspects of IT from development and operations to strategy and architecture and he has held CTO, CIO, CMO, and CEO roles. Throughout his career in technology, he has remained focused on what could be achieved in business terms and on business transformation. As well as various articles for academic and professional journals, he has authored two other books on IT management, and in 2013 was awarded a PhD for research into the management and governance of business technology. Other interests include motor sport and fencing. As a fencer and member of the Great Britain team, he has a gold medal from the 2016 European Veterans Team Championships.

List of Figures

List of Tables

The Need to Change the Way We Work 1

This chapter looks back at the way technology has been managed historically, the difficulties that this has created, and why it is necessary to rethink the problem as we look to the future. The material was mostly sourced from a review of academic literature which consists of peer-reviewed research findings by others in the broad area of the management and governance of business technology but it also contains some anecdotal evidence from what people in leadership roles are experiencing right now in what is a rapidly changing world.

Information technology (IT) has always been used by businesses to introduce greater efficiencies and create new commercial opportunities. Business re-engineering has been constant over the last six decades as the enabling technologies have evolved. At each phase this has not just yielded greater business efficiency and effectiveness benefits but also new business models in every business sector. In spite of our many successes, largely because the technologies themselves have moved on, in this chapter it is argued that the IT management methods have not. This has created problems for the IT professionals and left our IT leaders to defend the indefensible.

The first three sections of this introductory chapter look at what others have said, it considers 'What we know about our success rate', 'What we thought we knew about methods and standards', and 'What we should know about the management of business technology'. The remaining sections question the fundamentals of why we do things in a certain way, point to the further unintended consequences of our methods, and look to the future, why we must change the way we work and hence why the research that revealed the new model was necessary.

What We Know About Our Success Rate

What the last 50 or 60 years has taught us is that building IT business systems is inherently risky. In fact, the success rate has long been acknowledged to be around 35%, a figure first highlighted at a NATO conference in 1968. The comparable

D. Miller, *Managing Agile Business Technology*, Management for Professionals,
https://doi.org/10.1007/978-3-030-90598-9_1

success rate in the public sector has been reported to be more like 30%. These percentages are generally based on the number of objectives that have been met as a percentage of the total.

The reasons given by researchers and research groups reporting these statistics vary depending very much on the focus of the research undertaken. They include a lack of involvement of senior business managers, a weak business case, a need to better understand the business implications, more competent management of business integration, project management methods generally, project size, project planning, risk management, and estimating. Recent secondary research suggests that 35% continues to be a consistently reliable estimate of success rates in spite of the ever-increasing number of methodologies designed to fix the problem.

Business technology presents problems for its managers and its business stakeholders alike and high-profile investment failures are frequently reported in the media. These though are just the tip of the iceberg.

What becomes apparent to anyone entering the business technology profession is just how difficult it can be to get a definition of what the business requires, the complexity of some of the technical solutions, the difficulties associated with maintaining a viable technology capability, a realistic assessment of project progress, and the uncertainty of the eventual outcome from investment projects.

Failure is perhaps most often blamed on poor project management and there is little doubt that more qualitative analysis of project failure is necessary (Dalcher & Drevin, 2003) but anecdotally the most failures are probably caused by uncertainty and dysfunctionality.

Boehm claims that, based on his data from TRW, the estimating uncertainty reduces with each phase of project delivery (Boehm, 2008). This creates a 'cone of uncertainty', so named by McConnell (1998). By contrast, Little is strongly of the view that his data from Landmark Graphics over many years suggests that uncertainty is constant over the life of a project and this uncertainty must be managed (Little, 2006). Boehm claims that uncertainty is reduced through prototyping, simulating, modelling, benchmarking, reference checking, COTS (commercial off-the-shelf product) evaluation, and market trend analysis, but acknowledges that it is necessary to be agile enough to be able to adapt to the 'additional emergence of new sources of unpredictable change'. Thus, Boehm and Little may not be so far apart in their views—though neither has solved the problem.

Anecdotally we know that there are many causes of dysfunctionality to add to this uncertainty and as a consequence there is generally a poor relationship between the people delivering these systems and the business people which gives rise to much dissatisfaction.

This state of affairs continues today. Information Age published an article in March 2019 about a recent survey by Celonis (Baxter, 2019). According to the survey 44% of senior leaders believe their business transformation has been a waste of time and no less than 58% of business leaders admit that frontline employees are only involved in transformation projects because middle management or consultants tell them which changes to make. The report also states that no less than 45% of

C-suite executives admit they do not know where to start when developing their transformation strategy.

The problem is a global one and with project success rates remaining stubbornly low what focuses the attention of business management is the consequential cost of waste; it is suggested that globally some $2.5 trillion to $3 trillion (Krigsman, 2012) is at risk annually and rising. The successes of the semiconductor industry have probably sheltered the IT community at large from its poor success rate in IT delivery. If there is any slowdown in the price performance improvement that we have become used to, it is likely that businesses will be encouraged to look more closely at the cost of waste associated with IT investments.

What We Thought We Knew About Methods and Standards

Conventional wisdom states that our methods and standards have evolved over many years and they are designed to reduce the risk of failure by applying best practice across not just the project management processes but across all of the delivery activities. But is this the case? In an age where technology is embedded within almost every critical business process and yet where the failure of technology investments still appears to be statistically more likely than not, business people can be forgiven for thinking that the management of the delivery of technology is lagging behind the technology itself. It is not surprising that there has been a focus on delivery best practice, and hence methodologies, but to what end?

With the spotlight on IT delivery, failure to deliver is not just the concern of the CIO. Business managers will be concerned about the prospect of the failure of an IT investment that it needs in order to remain competitive. The CEO will be concerned that the business may be put at risk by making large investments in such technology. The CIO may find it necessary to provide assurances that any risks are manageable—so where do these risks lie?

The methods, standards, and practice have dictated that there are four key delivery components: the requirements specification is the blueprint for the design, architectures provide the context, change management maintains the currency, and standards ensure fitness for purpose. Examining the nature of the current IT delivery model, i.e. as exemplified by current practices, it consists of processes for business analysis, defining the requirement, architectures/architectural alignment, solution design, development, testing and integration, the deployment of the solution, the management of operational services, measurement of operational performance, application maintenance, and program and project management, security and regulation (legal compliance), etc. There is one delivery model addressing the scope of these activities, a bricolage of IT management tools, and an infinite number of implementation variations.

The IT management tools don't neatly map to the above processes, nor do they slot together in a predetermined way with clear interfaces. The cocktail or mixture of tools to be used for any particular implementation, we'll call it the modus operandi, will be decided by the management team based on any house rules, the task and their

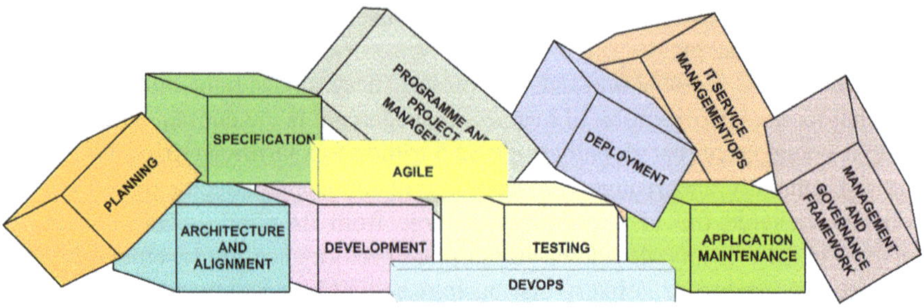

Fig. 1.1 A bricolage of IT methods

preferences for agile methods/not, their perceived needs, and their skill levels across the management tool kits. There has been a dependence on methods and yet there are now so many variants and extensions that we 'mix and match', 'adopt and adapt', or 'pick and mix' to meet needs and this is just as true when using agile approaches (Fig. 1.1).

The IT management tool kits have become the focus of considerable investment. Some of them are so embedded in the culture they often define the functional and commercial relationship between TSPs and the business. Judging the outcome from an investment in technology can be problematic, especially when success rates are so low, and many people in the industry found it easier to demonstrate compliance with method, i.e. best practice, as opposed to ensuring that the outcome met the business need. The IT methods and management tools themselves are blamed by many for the frequent failures quoting their overlap and a lack of standardisation. There is a generally accepted view that IT delivery is failing to meet business expectations, and it has even been said that a development outcome may be a superb technological success but a total failure for the business.

It has not been possible to find any estimates of the spend globally on training, certification, and accreditation associated with IT methods or estimates of the benefits of using methods and standards to offset the cost of waste of $3 trillion but consider that in a sample of 40 projects there was no correlation between a successful business outcome and the use of the frequently applied IT management methods. This prompts the questions: what control can we rely on these management methods providing, and at what cost?

What We Should Know About the Management of Business Technology

Whilst the focus for methodologies has been the delivery model, it is clear from the literature that the macro management of business technology has also always been a weakness. Alignment and governance are the key concepts at this level and these have largely escaped scrutiny.

It was probably Zachman who first conceptualised architecture (Zachman, 1978). Many have since recognised the parallels between this basic architectural framework and the need for business and IT alignment. The uncertainty of the alignment process was first noted by researchers at MIT (Henderson & Venkatraman, 1989) and later the lack of business and IT alignment was linked with IT investment failure (Velitchkov, 2008). Some have been of the view that the solution to achieving alignment is to embed the IT strategy and business requirements into the Enterprise Architecture (EA). This has prompted concerns that such EA concepts are being pushed beyond what can be justified and that, taken too far, can adversely affect the TSP's ability to deliver useful business functionality to end users in a timely manner (Kemp & McManus, 2009).

It is not just the uncertainty that is limiting the usefulness of the EA approach but the growing number of variables, even within the architectural, planning, and procedural aspects which are collectively referred to as the intellectual dimensions. It is now widely recognised that alignment also needs to consider the 'abstract', i.e. the social and cultural dimensions. The work identifying these abstract dimensions is in its infancy. The scale and scope of the research into these aspects to date has been limited with no agreement on the dimensions or how these are to be measured other than an acknowledgement that the non-alignment of the abstract dimensions can be a cause of failure and that more research was needed.

Perhaps another reason for failure is that research has not clearly defined alignment, there was still little agreement over what alignment is, where it is most likely to occur, how it can be achieved, or how it should be measured. What we have relies instead on tautological definitions (Maes et al., 2000) and a tendency to use theoretical and artificial constructs for measurement. The literature often focuses on justifying and trying to define business-IT alignment but it doesn't actually explain how to do it other than offering simple rules of thumb and observing that alignment is generally considered to be a continuous process embracing both the strategic and tactical levels.

It has been recognised that in practice the real world provides a more complex and fuzzier situation in which alignment is not as straightforward as implied by the methodologies perhaps because of the uncertainty of the process (Henderson & Venkatraman, 1989), because business strategies are often unclear (Silvius, 2007), and because business goals are frequently high level and generic (Grembergen et al., 2005).

If alignment is a key output from a delivery model, it would probably not be possible to design it from the literature. If business strategies are often unclear and alignment cannot be defined or measured, then it is not surprising that in surveys so many CIOs report alignment as poor or fair and, as another researcher observes, it continues to be amongst the top 10 issues for many organisations (Cumps et al., 2006).

Governance is often mentioned simultaneously with alignment but the relationship is not explored and, like alignment, governance is similarly not well defined. There is a gulf between the IT controls at the macro (leadership) and micro (operational) levels (Ahituv et al., 1999), and there is little consensus as to how the gulf must be filled.

At the micromanagement level IT governance is seen by the IT people as IT management as defined by the adopted IT methodology. In these situations, the business people may see IT governance as being imposed by the IT organisation (DeHaes & Grembergen, 2005) and, because IT governance is often explained in terms of the IT management mechanisms used, this is failing most businesses (DeHaes & Grembergen, 2009). It is worth adding that the ISO standard for the corporate governance of information technology is focused on the outputs of managing 'ICT projects' and 'ICT operations' so it is only offering a micromanagement solution.

At the macro level it has been said that there is no one solution for the governance of business technology that fits all boards of directors (Nolan & McFarlan, 2005); the suggestion here is that solutions will vary widely across organisations. Nolan and McFarlan suggest a framework which classifies a business according to how it uses IT and a calendar for the IT governance committee in which the topics deemed necessary for that business are identified and scheduled to be discussed quarterly, annually, or 'as required'. Many management teams will see this as being less than helpful and it can't be reconciled with the need for either agility or continuous alignment.

It is necessary to reconsider how the governance of business technology can be improved. Whilst putting to one side their various solution propositions, many people have identified the need for governance to involve the business more, to address a broader spectrum of business needs, to connect with business more effectively, and at more levels, and for this to be a continuous process (Campbell et al., 2005; Kang et al., 2010; Khaiata & Zualkernan, 2009; Schepers et al., 2008; Silva & Chaix, 2008; Wegmann et al., 2007). This is a powerful message, and a solution is needed not just for this world, but also for a future self-adaptive and self-managing world where a more immediate, response will be demanded.

Much is written about the failure of IT management (Cooke et al., 2001; Hughes, 2008; Poba-Nzaou et al., 2008; Sauer & Cuthbertson, 2003; Smith & Booth, 2000; Whittaker, 1997) and especially where methods are mandated (Ben-Menachem & Gelbard, 2002; Bicknell, 2012; Little, 2006; Velitchkov, 2008; Winniford et al., 2009). Whilst the problem is widely recognised no workable solutions are offered and perhaps this is why some businesses see IT (or more correctly the TSP) as an obstacle to be overcome. This unfortunate situation can only get worse unless we change the way we work.

Questioning the Fundamentals

If the management and the delivery processes are failing business in some way, we should look at the fundamentals to understand what is going on. Our methods are deterministic so the requirement is set externally to the framework we choose to use. This is because most IT methods and standards were originally explicitly or

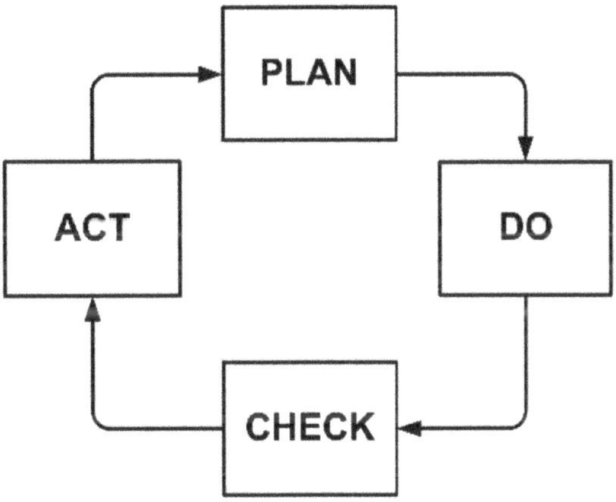

Fig. 1.2 The Shewhart or Deming Cycle

implicitly based on the Shewhart Cycle, sometimes referred to as the Plan, Do, Check, Act (PDCA) framework. It's the reason that the first question that TSPs ask of business people is likely to be, 'What do you want?'. It is also the framework applied by Deming to early industrial processes most famously in the motor industry, see Fig. 1.2. The significance of this determinism will become apparent in the next chapter but for now we can consider its effect on the way we work.

Being manufacturing based, the 'Deming Management Method' is product and production oriented (Rungtusanatham et al., 2003). The Deming focus was towards quantitative measurement requiring targets to be achieved (Petersen, 1999). PDCA became the bedrock of Total Quality Management (Knouse et al., 2009) and was featured explicitly or implicitly in IT methods, standards, software engineering, and Six Sigma (Pande et al., 2000). Though the Shewhart Cycle may be interpreted with minor variations according to the particular application it is always used deterministically where processes of the type 'Planning', 'Doing', and 'Checking' are focused on achieving a planned outcome. If this is a simple task (such as an operation on an assembly line), the Shewhart Cycle needs no modification but for IT project management methods where there are multiple phases each phase of the project could be a PDCA cycle as above or an alternative interpretation could group the 'DO' phases as in Fig. 1.3, where each 'DO', in agile terms, may be a sprint. CHECK here may perhaps refer to the post implementation review and ACT to the process of feeding back to the next cycle of planning. Anecdotally we know that both of these latter processes are often given short shrift or even omitted completely.

Where there is more emphasis on planning, e.g. for management and governance purposes especially in large and complex enterprises, PDCA is still used as a basis of the methods but adapted as, for example, in Fig. 1.4. Here the need is often to coordinate different subjects that have different planning cycles, e.g. the business planning cycle (on the left) and the technology planning cycle (on the right). ACT in this model is the process of gathering the necessary information. PLAN is about updating the strategic plans and dovetailing business and IT priorities for requirements planning, portfolio management, architecture development, and

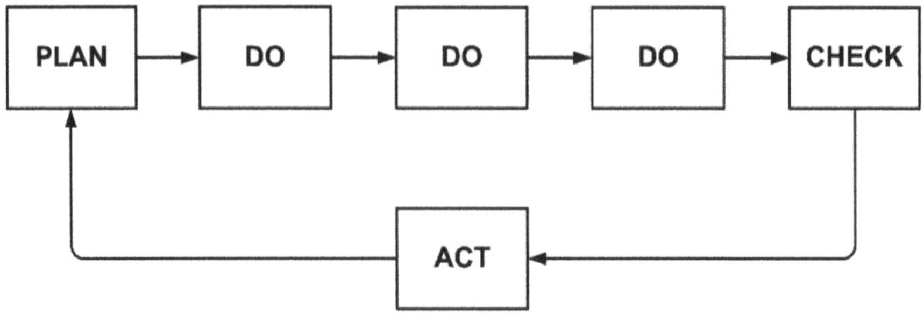

Fig. 1.3 Project management vs The Shewhart or Deming Cycle

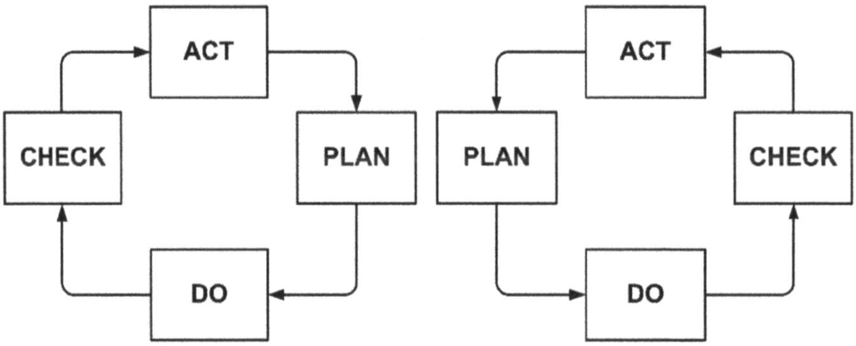

Fig. 1.4 Dynamics of demand and supply vs The Shewhart or Deming Cycle

supplier capability. DO will involve change programs and infrastructure projects, and CHECK will be the assessments of what is delivered.

It is easy to see how, when the traditional methods combine different types of framework, it could take a long time to do almost anything. Time can also be the enemy of the certainty of the outcome, i.e. the longer something takes, the more likely the outcome is likely to be in jeopardy as a consequence of both internal and external factors.

These traditional methods and standards are enforced using process maturity techniques such as CMMI and ISO 15504. Thus, the IT industry has created a situation where process is the focus, where accreditation is only possible if the process is followed, but where success rates are low perhaps because of external uncertainty, dysfunctionality, the variable nature of the delivery model or flaws in the delivery model itself, or a lack of definition of the alignment and governance process. Nowhere in the literature does it suggest that the current methods contribute to failure—but given the scale of training, certification, and accreditation available, it has often been said at senior levels, and in government circles, that IT people are well qualified in acceptable ways of failure. It's not clear whether this is a criticism of the methods, the training, or an acknowledgement of the nature of the task.

Process maturity is further emphasised by over simplistic mantras such as 'people, process and technology' that focus entirely on value *delivery*, i.e. as Boehm says there is little space for value creation during value-based software engineering (Boehm, 2003). The software engineering body of knowledge is focused on the construction of technology solutions. It is geared up to supply features not value. The need and eventual specification may have evolved from a strategy study but we know that kind of business-focused initiative is often a one-off which quickly gets out of date. Indeed, incorrect or incomplete requirements specifications are commonplace. There is no mention of timing; a late solution may have no value. Oversimplified mantras are likely to lead to more uncertainty and dysfunctionality.

The stubbornly low success rate suggests that our traditional methods based upon the product and production paradigm are either incorrect or incomplete. By contrast, service can be considered more as the co-creation of value, (Spohrer & Maglio, 2008). The service economy is based on value creation and the technology services provider, whilst recognising that there are still product and production issues to be managed, must ensure that meeting the many needs of the business and consumer is the main quality driver for IT services (Vouk, 2008). Under these circumstances, and without some form of validation from the business, it's easy for everyone involved, including TSPs, to delude themselves that they are on the right path (Ries, 2011). To date the industry has had no reliable business validation process. Could this be the key to our alignment and governance needs?

We have essentially always worked with information technology in the same way. Although much has been achieved the uncertainties associated with defining the requirement and managing the IT investment in an increasingly complex, ambiguous, and dynamic context are often too great. Added to this, service, value, and agility are now seen as the management challenge and yet over the lifetime of the current IT delivery model this was not the case. The nature of the management challenge has changed dramatically and further change is still taking place. We have moved from a predominantly production economy to a service one, and we have seen technology play an increasingly important role not just in making business more efficient but also in actually creating markets and facilitating opportunity. Accordingly, we need to change the way we work.

The Unintended Consequences of Past Methods

Regardless of the problems described in the earlier sections of this chapter we have muddled through and achieved quite a lot. The way we have worked has, though, resulted in further unintended consequences.

The first is a flawed storyline. The current story line involves process which often takes too long, objectives are missed, and external events can sometimes overtake the projects before they are complete with the result that TSPs are always playing catch-up not knowing where the next requirement is going to emerge. This is an obvious flaw that needs to be fixed and it can only be changed by changing the way we work.

The second is that the management methods have resulted in a further separation of the business and IT roles. In many cases the IT roles will even have been outsourced to a TSP or even moved offshore. There is no doubt that this separation has further exacerbated the already damaged language barrier between business and IT people. Superficially and anecdotally, business people complain about the use of jargon and technology people tell us that business people never know what they want. In the next chapter we will discover why both views are often correct. Furthermore, the best ideas often come from unlikely places so it's difficult to know who best to speak with to determine the requirements, and, to complicate matters further, it has been said that there is often a lack of access to real stakeholders during design. Few business people are capable of fully defining their own requirements and service levels to the satisfaction of the TSP. It is now common practice for a TSP to specify both the functional and technical aspects of a requirement based on some input from business people but with the inevitable risk of a misunderstanding. Even when we introduce more rigour to the management process in the form of configuration management, validation, and verification techniques, they can never be absolute, i.e. the information we gather is never complete, but we don't know where the gaps in our knowledge lie, and things will change, but we don't know what will change. To eliminate any problems associated with language, business and IT people must have a more effective relationship where understanding is guaranteed or at least the chances of a misunderstanding are greatly reduced. More than merely communicate, we need business and technology people to collaborate.

The third unintended consequence is that our management methods have failed to adapt to the real world which has become more complex, ambiguous, and dynamic. Regardless of the modus operandi chosen for a given project or by a particular organisation the mantra demands static constructs, i.e. the requirement and the architecture in which it sits must be fixed or stable. In fact, this is incorrect because uncertainty always prevails. Change can be sudden and have a major impact such that effort expended, rather serving any useful purpose, may have to be written off. Static complexity (e.g. as embodied within the enterprise architecture) can often introduce those tautological definitions for alignment, including the social dimensions and the many implementation level extensions, e.g. for UML, RACI, TOGAF, which together add such complexity. Linking the conceptual EAModel and the implementation level Configuration Management Database (CMDB) is a very complex problem and one often avoided in the absence of total automation. Agile is often used to try to offset time lost caused by reworking static complexity. Both professionals and academics raise questions about agile, in particular how agile IT methods fit with big business, business agility, and even the Agile Manifesto. Many people have commented on the difficulties of introducing agile methods to non-agile businesses—and conversely just how do businesses become agile when the methods are seen as the greatest inhibitor to change? This suggests the eventual breakdown of methods but because of the investment in them the process may take a long time to play out. We are told that in the alignment process we must be prepared to continuously make adaptations (Henderson & Venkatraman, 1993) but the static constructs that our methods rely upon do not facilitate this way of working.

We need to find a way of changing the storyline so we are not having to play catch-up, we need to bring business and IT together again for the good of the business, and find a way of managing a complex, dynamic, and sometimes ambiguous world without an over-reliance on static constructs.

Looking to the Future

The world is not going to go back to what it was because actually it never was how we then saw it. Communicating what has to be done was never that simple and service, value, and agility were actually always important. Also, technology continues to advance with digital transformation, the Internet of Things (IoT), Big Data, Artificial Intelligence (AI), multi-agent systems, Quantum Computing, self-adaptive and self-managing systems, nanotechnologies, advanced materials, advanced manufacturing and processing, and things I've forgotten or we don't yet know about will introduce further transformations.

The business world has always been in a state of flux but if service, value, and agility are the challenge then new ways of doing business and new technology will be seen as the means of delivery. Recent observable and self-evident trends have forced us to rethink delivery, and in combination, are compelling. Commoditisation, e.g. exemplified by cloud computing, has removed many of the capital constraints associated with satisfying new business needs. Our connected world, the IoT, and the consumerisation of corporate business systems using a mobile interface, e.g. an app on a smart phone, saves time and improves data quality and decision making whilst freeing up our people from their desks. Collectively these developments allow internal and external stakeholders to simultaneously and willingly become both a consumer and a configurator of services (Sharif, 2010). As a consequence, the number and size of communities supported by our systems has increased. What hinders progress as scale and complexity grow is our ability to manage service, value, and agility to all of these different communities in a balanced manner. As the current storyline dictates, playing catch-up on this scale, not knowing where the next requirement is going to emerge, is not practical. We need to change the way we work, e.g. by orchestrating the many communities to ensure that they receive new or modified solutions as needed, i.e. at speed.

There are many other reasons for the increasing complexity, e.g. the technology itself, globalisation, distributed and extended enterprises, legacy systems, the growth in cybercrime, and increasing regulation, etc. The quick fix is to do what the contract says, i.e. 'follow the money', but if complexity is reaching the limits of human capability (Kephart & Chess, 2003) and, as Boehm was reported in the Irish Times as saying, 'we live in an increasingly complex world with new failure modes and new risks' (Lillington, 2005), we need something better. We have seen that we have always struggled with alignment and governance, value creation, and even value delivery. To manage in a more complex world we need to change the way we work.

The separation of the business and IT roles has become problematic. We see from lists of top CIO issues that CIOs are increasingly being asked to work on business innovation and the need to find new business revenue streams, whilst at the same time business leaders are being tasked to consider uncertainty, when to embrace change, complexity, the management of information and data analytics, balancing the interests of the many different stakeholders, strategic thinking, identifying mega-trends, and problem solving, customer service and maintaining reputation. These issues increasingly necessitate close collaboration between the functions of the business and the TSP. We are certainly seeing a need for a common language between the business and technology aspects to avoid the necessity for translation, i.e. we bring together the syntax (rules for construction) and knowledge of the semantics (meaning of the words) to achieve cognition (understanding) as in the Chinese Room problem (Searle, 1990). Maybe it's more than just collaboration, with both parties often citing competencies and recruitment as key issues maybe we are seeing the need for hybrid managers, i.e. tech-savvy business managers. Failure to bring the parties together will have negative consequences. Disruptive technologies, new competition, and technology investment failures can lead to distrust, reduce expectation, lead to a reassessment of the risks associated with IT investment, of investment being limited (especially during times of austerity), or prompt discussions about the future of the IT supply organisation. Any IT credibility gap can lead to a reluctance to invest further which could have a negative impact on the commercial performance of the business. To encourage collaboration, we need to change the way we work.

The manner in which we construct solutions is changing. Alongside agile methods, web services and SOA are often used to define units of functionality that can be flexibly modified or redeployed to support different business contexts (Diaconita, 2011) enabling a multi-dimensional alignment strategy (Chen, 2008), i.e. architecture, governance, and communication, and the resource-based view theory perspective (Wernerfelt, 1995). Value is created by integrating these resources dynamically and alignment can become a continuous process. Service orientation 'must be capable of handling change requests that may arrive at any time' and so 'SOA governance is not a process but a matter of continuously aligning strategic goals, new tactical opportunities, and to use gained experiences' (Schepers et al., 2008). Our traditional management methods don't work like this and so, to maximise the opportunity this presents to the business, and manage it effectively, we need to change the way we work. Multi-agent systems build on this concept and again require a different approach.

Longer term, a more flexible and agile approach to business and technology management is required to enable business to react quickly to whatever risks or opportunities present themselves (Angell & Smithson, 1990). This will inevitably lead to the eventual automation of the management of business technology. This has been recognised for some time but it's not within the capability of the current IT delivery model. Automation today is restricted to autonomic computing in service management (McBride, 2002) to assist in the recovery from operational failure, e.g. server, network, hardware component failure, or software update failure. Automation is not yet capable of adapting an application to meet some new business requirement. When truly self-adaptive or self-managing systems are able to constantly respond to new business requirements defining and testing the requirement will be difficult if not impossible except at run time (Kephart & Chess, 2003). If this approach is to prevail, success rates must be very much greater than those experienced today and they will need to be maintained continuously.

Many discussions about the future of computing raise the prospect of computers with the ability to think for themselves, and bio-computing, the merging of biological processes and computers. The combination of these technologies opens up the prospect of computers that can undertake the fundamental cognitive processes of natural intelligence such as learning, thinking, formal inference, and perception processes, i.e. cognitive informatics or novel information processing systems that think and feel (Wang, 2006). Boehm when talking about the dangers of automation introducing new failure modes for computing warns about the loss of human primacy and even the overpowerment of humans (Lillington, 2005). Will humans become subservient to computer algorithms or will we see the emergence of techno-humanism or transhumanism (Harari, 2011)? These technologies present new challenges for business technology management. How is automation to be managed to ensure that the outcomes are compatible with the values of the business and its management and the ethics of society at large? Our current delivery model and management methods are at the limits of what they are capable of doing. To both introduce this level of automation, and to manage it, we need to change the way we work.

We have so far, more or less, got by with our old methods. Nevertheless, this chapter has listed many reasons why we should change the way we work and why we need improved management methods for the emerging and future worlds as we perceive them. We need new ideas and we know that little original research has been carried out in this area. Anecdotally we would expect future technology management methods to place the business, or people, at its centre and to react quickly to whatever risks or opportunities present themselves. Methods need to simplify, facilitate, and even automate the introduction of new technologies, services, and solutions. We have recognised the need to work more quickly but, in many cases, business is still too remote from the technology. We must look to automate the technology management and governance process itself such that it is more integrated, agile, balanced, controlled, and successful. To take this necessary step forward more original research was necessary.

Chapter 2, Breaking Through, is about the discoveries resulting from one such research project, most notably the Business and Technology Relationship Model (BTRM). The BTRM is described in detail. The chapter explains the BTRM's significance in terms of both the management and governance of business technology and how it changes the storyline for both the business and the TSP. The remaining chapters describe how by implementing a conceptual system called Active Management and Governance, based on this model, that we can change the way we work and so manage business technology, and business, in ways we hadn't previously thought possible.

References

Ahituv, N., Zviran, M., & Glezer, C. (1999). Top management toolbox for managing corporate IT. *Communications of the ACM, 42*(4), 7.

Angell, I. O., & Smithson, S. (1990). Managing information technology: A crisis of confidence? *European Management Journal, 8*(1), 27–36.

Baxter, M. (2019, March 21). Business leaders disillusioned with business transformation, research report. *Information Age*. Retrieved from https://story.celonis.com/square-one-research/

Ben-Menachem, M., & Gelbard, R. (2002). Integrated IT management tool kit. *Communications of the ACM, 45*(4), 7.

Bicknell, D. (2012). Cabinet office promises to challenge 'culture of secrecy' on IT projects. *The Guardian*. Retrieved from http://www.guardian.co.uk/government-computing-network/2012/jun/13/transparency-projects-it-gateway-review-cabinet-office

Boehm, B. (2003). Value-based software engineering. *ACM Software Engineering Notes, 28*(2), 12.

Boehm, B. (2008). Making a difference in the software century. *Computer, 2008*, 32–38.

Campbell, B., Kay, R., & Avison, D. (2005). Strategic alignment: a practitioner's perspective. *Journal of Enterprise Information Management, 18*(6), 653–664. https://doi.org/10.1108/17410390510628364

Chen, H.-M. (2008, January 22). *Towards service engineering: Service orientation and business-IT alignment.* Paper presented at the 41st Annual International Conference on System Sciences Hawaii.

Cooke, D., Gelman, L., & Peterson, W. J. (2001). *ERP trends*. The Conference Board.

Cumps, B., Viaene, S., & Dedene, G. (2006). Managing for better business-IT alignment. *IT pro, Sept/Oct 2006*.

Dalcher, D., & Drevin, L. (2003). *Learning from information systems failures by using narrative and ante-narrative methods*. Paper presented at the South African Institute for Computer Scientists and Information Technologists.

DeHaes, S., & Grembergen, W. V. (2005). *IT governance structures, processes and relational mechanisms: Achieving IT/business alignment in a major Belgian financial group*. Paper presented at the 38th Hawaii International Conference on System Sciences, Hawaii.

DeHaes, S., & Grembergen, W. V. (2009). An exploratory study into IT governance implementations and its impact on business/IT alignment. *Information Systems Management, 26*, 123–137.

Diaconita, V. (2011). Considerations regarding designing and administrating SOA solutions. *Database Systems Journal, 11*(2), 10.

Grembergen, W. V., DeHaes, S., & Moons, J. (2005). Linking business goals to IT goals and Cobit processes. *Information Systems Control Journal, 4*, 18–22.

Harari, Y. N. (2011). *Sapiens: A brief history of humankind*. Harper.

Henderson, J. C., & Venkatraman, N. (1989). *Strategic alignment: A framework for strategic information technology management Sloan working papers*. Massachusets Institute of Technology.

Henderson, J. C., & Venkatraman, N. (1993). Strategic alignment: Leveraging information technology for transforming organisations. *IBM Systems Journal, 32*(1), 4–16.

Hughes, B. (2008). *Exploiting IT for business benefit* (1st ed.). The British Computer Society.

Kang, D., Lee, J., Choi, S., & Kim, K. (2010). An ontology-based enterprise architecture. *Expert Systems with Applications, 37*, 1456–1464. https://doi.org/10.1016/j.eswa.2009.06.073

Kemp, P., & McManus, J. (2009). Opinion: Enterprise architecture - Dogmatic and over-ambitious? *ITNOW, 51*, 20–21.

Kephart, J. O., & Chess, D. M. (2003). *The vision of autonomic computing computer* (pp. 41–50). IEEE Computer Society.

Khaiata, M., & Zualkernan, I. A. (2009). A simple instrument to measure IT-business alignment maturity. *Information Systems Management, 26*, 138–152.

Knouse, S. B., Carson, P. P., Carson, K. D., & Heady, R. B. (2009). Improve constantly and forever the influence of W. Edwards Deming into the twenty-first century. *The TQM Journal, 21*(5), 449–461. https://doi.org/10.1108/17542730910983371

Krigsman, M. (2012). Worldwide cost of IT failure (revisited): $3 trillion *ZDNet* (pp. blog post talking about a rationale for calculating the global cost of IT failure prompted by flawed estimates by Roger sessions of $6 trillion).

Lillington, K. (2005). Visionary professor takes a look at the wild-card future of computer science, *Irish Times*, pp. 7–7.

Little, T. (2006). Schedule estimation and uncertainty surrounding the cone of uncertainty. *IEEE Software, May/June 2006.*

Maes, R., Rijsenbrij, D., Truijens, O., & Goedvolk, H. (2000). *Redefining business-IT alignment through a unified framework.* Paper presented at the Landelijk Architectuur Congres 2000, Amsterdam.

McBride, N. (2002). Autonomic computing and the future of IT service management. *Journal of the IT Service Management Forum, 2002*, 4.

McConnell, S. (1998). *Software project survival guide: How to be sure your first important project isn't your last.* Microsoft Press.

Nolan, R., & McFarlan, W. (2005, October 11). Information technology and the Board of Directors. *Harvard Business Review.*

Pande, P. S., Neuman, R. P., & Cavanagh, R. R. (2000). *The six sigma way.* McGraw-Hill.

Petersen, P. B. (1999). Total quality management and the Deming approach to quality management. *Journal of Management History, 5*(8), 468–488.

Poba-Nzaou, P., Raymond, L., & Fabi, B. (2008). Adoption and risk of ERP systems in manufacturing SMEs: a positivist case study. [Research paper]. *Business Process Management Journal, 14*(4), 20. https://doi.org/10.1108/14637150810888064

Ries, E. (2011). *The lean startup - How constant innovation creates radically successful businesses.* Penguin Books (USA) ed. Crown Publishing Group.

Rungtusanatham, M., Ogden, J. A., & Wu, B. (2003). Advancing theory development in total quality management: A "Deming management method" perspective. *International Journal of Operations & Production Management, 23*(8), 918–936.

Sauer, C., & Cuthbertson, C. (2003). *The state of IT Project Management in the UK 2002-2003* (p. 82). Templeton college, University of Oxford.

Schepers, T. G. J., Iacob, M. E., & Van Eck, P. A. T. (2008, March 16–20). *A lifecycle approach to SOA governance.* Paper presented at the SAC'08, Fortaleza, Ceara, Brazil.

Searle, J. (1990). Is the Brain's mind a computer program? *Scientific American, 1990*, 6.

Sharif, A. (2010). It's written in the cloud: The hype and promise of cloud computing. [viewpoint]. *Journal of Enterprise Information, 123*(2), 4. https://doi.org/10.1108/17410391011019732

Silva, E., & Chaix, Y. (2008). *Business and IT governance alignment simulation essay on a business process and IT service model.* Paper presented at the proceedings of the 41st Annual Hawaii International Conference on System Sciences (HICSS 2008), Hawaii.

Silvius, A. J. G. (2007, 29 January). *Business & IT alignment in theory and practice.* Paper presented at the 40th Annual Hawaii International Conference on System Sciences, Waikoloa, HI.

Smith, T., & Booth, R. (2000). *IT project failures costly, TechRepublic/Gartner study finds.* Retrieved March 30, 2011, from http://www.techrepublic.com/article/it-project-failures-costly-techrepublicgartner-study-finds/1062043

Spohrer, J., & Maglio, P. P. (2008). *The emergence of service science: Toward systematic service innovations to accelerate co-creation of value.*

Velitchkov, I. (2008). *Integration of IT strategy and enterprise architecture models.* Paper presented at the International Conference on Computer Systems and Technologies - CompSysTech'08.

Vouk, M. A. (2008). Cloud computing – Issues, research and implementations. *Journal of Computing and Information Technology, 16*(4), 12.

Wang, Y. (2006). *Cognitive informatics: Towards future generation computers that think and feel.* Paper presented at the 5th IEEE Int. Conf. On Cognitive Informatics (ICCI'06), Beijing, China.

Wegmann, A., Regev, G., Rychkova, I., Le, L.-S., & Julia, P. (2007). *Business and IT alignment with SEAM for enterprise architecture*. 11th IEEE International Enterprise Distributed Object Computing Conference (EDOC 2007), 111.

Wernerfelt, B. (1995). The resource-based view of the firm: Ten years after. *Strategic Management Journal, 16*(3), 171–174.

Whittaker, B. (1997). What went wrong? Unsuccessful information technology projects [research paper]. *Information Management & Computer Security, 7*(1), 7. https://doi.org/10.1108/09685229910255160

Winniford, D. M., Conger, S., & Erickson-Harris, L. (2009). Confusion in the ranks: IT service management practice and terminology. *Information Systems Management, 26*, 12. https://doi.org/10.1080/10580530902797532

Zachman, J. A. (1978). The information systems management system: A framework for planning. *Data Base, 1978*, 6.

2 Breaking Through

This chapter describes the Business and Technology Relationship Model (BTRM) and other key findings from the largest research project of its kind using high quality primary research data from three large complex global businesses. It was instigated as a result of the issues raised in Chap. 1 and the findings provide an insight into why existing methods can fail and how that relationship between business and technology might be better managed.

The first section, 'An introduction to the BTRM' describes the model itself, section two 'Talking about different things' explains what was discovered about the communications problems often remarked upon between business people and IT people, section three 'Spectrum of Business Requirement' covers how business people define their various needs in different ways and at different times, section four 'TBE: a measure of service and indicator of Change' explains the importance of managing the total business experience, section five 'Blind spots causing failure' illustrates the potential weaknesses of current methods in the light of the findings, section six 'The BTRM as a governance framework' shows how two frequently raised issues for the CIO, alignment and governance, are addressed by the BTRM. The final section explains how the BTRM fundamentally changes the storyline thus introducing Chap. 3 which describes the conceptual system 'Active Management and Governance', based on the BTRM, which explains how this storyline might be implemented and exploited.

An Introduction to the BTRM

Originally written up as being the BITRM (The Business and IT Relationship Model) the post research development showed that it can be more broadly applied to all business technologies, hence the change of name to BTRM. It visualises the relationship between business and the technology it uses and is supported by an ontology that would help to support any kind of implementation.

D. Miller, *Managing Agile Business Technology*, Management for Professionals,
https://doi.org/10.1007/978-3-030-90598-9_2

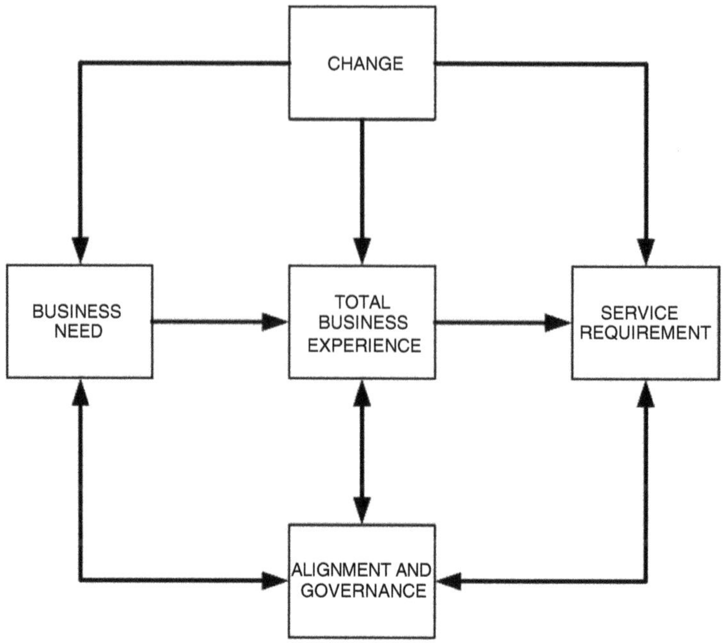

Fig. 2.1 The Business and Technology Relationship Model

From Fig. 2.1, we can see that the key categories of the BTRM are Change, Business Need, Total Business Experience (TBE), Service Requirement, and Alignment and Governance.

Change is a widely used term. We can't manage Change but if we don't know about Change, we can't properly respond to it. Change is of critical importance and the research shows that it is rarely communicated promptly, if at all. Instead, we try to manage change (in fact we manage requests for change as they arise, often placing them in a queue) instead of managing our response to Change as the BTRM suggests. At an early stage a distinction was made between an expected or actual event, typically external, that prompts a new Business Need, and this was called a Change, and the response of the business and technology people to that Change. The response will need to be managed and could involve the realignment of products, technology and other services, people, business processes, architectures, etc. Change can be considered to drive the Business Need, the TBE, and the Service Requirement. The Alignment Task is the response. Thus, the BTRM tells us that we don't manage Change but we respond to it. These categories and their relationship to one another will now be described in a little more detail. In addition, the reader may wish to refer to the appendix for further detail where all of the fields of all of the categories and their definitions have been documented.

Change is anything which changes the operating environment of the business that could represent an opportunity or threat. Examples may include something imposed by shareholders, changed trading circumstances, competitor activity, a downturn in the economy, or something new and relevant in terms of a technology, a failure of the technology of some kind or the services relating to it, or a combination of these. Change can be explicit and give rise to a new Business Need, a revision of our TBE (i.e. how we see the adequacy of our technology products and services in the light of the Change and the New Business Need), and Service Requirements (i.e. what services are thought to be required to start to address the issue being raised). Change is frequent, and as Anakin Skywalker in Star Wars was told when counselled by Shmi, 'You can't stop change any more than you can stop the suns from setting' (Brooks, 2000), however, many would like it to be as predictable.

Business Need is a new or changed business context, business strategy and objectives, and the resultant Needs Gap. This may have arisen as a consequence of a Change or it may simply be an efficiency or effectiveness improvement.

Total Business Experience (TBE) is the business perception of the suitability of the business technology products and services at that point in time. Assessments can go up and down and will reflect the business experience when using the products or services. If a Change is being indicated, it may be that the products or services are now unsatisfactory in some way.

Service Requirement is a request for services thought to be required. There are five broad dimensions of service request: Overall Service Management, Business Planning, Application/Digital Services, Transition Services, and Infrastructure Services.

Each can arise independently, and in any order, but over time links will be formed between these categories, i.e. a Change can give rise to a new Business Need, which in turn can give rise to a revised TBE, and this leads to a need for some form of service as expressed by a Service Requirement. When linked in this way this is referred to as the 'Causal Chain', see Fig. 2.2. If there is no obvious Causal Chain, the different requirements are referred to as a 'Spectrum of Business Requirement' which is discussed in a later section of this chapter. It is clear that when the causal links are forgotten or unclear the full significance of what is needed can be hidden.

Alignment and Governance is where an intervention in the form of an Alignment Task is initiated in response to elements of the Causal Chain. Note that both Change and the response to Change are both within the scope of the model. This provides important opportunities in the way the relationship is managed because the sooner Change is detected, the sooner the impact can be assessed, and the sooner a response can be delivered.

A final point that has future significance is that the BTRM has heuristic, deterministic, and control characteristics making process based on this model potential candidates for automation.

Talking About Different Things

Chapter 1 mentioned that language has long been a problem between business and IT people, i.e. the use of the jargon by IT people and business people never knowing what they want. The research process used Grounded Theory Method (GTM) which

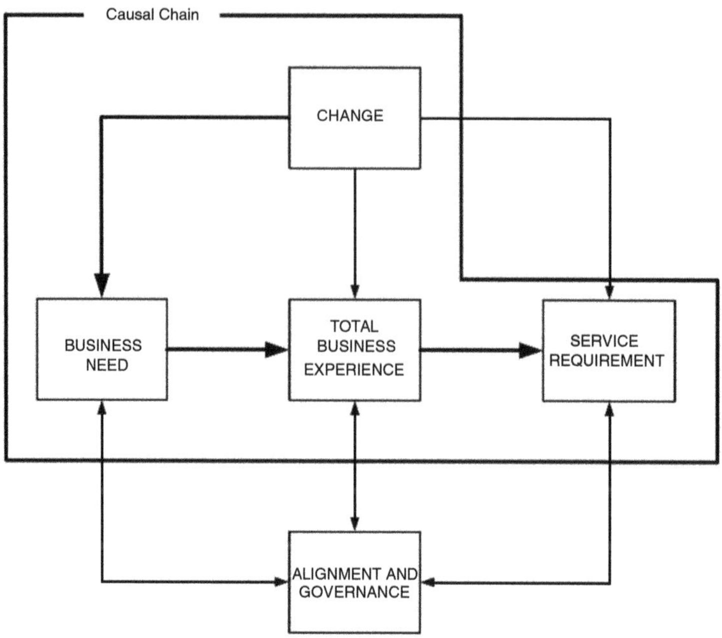

Fig. 2.2 The Causal Chain

forensically analyses what people are talking about and categorises the topics being discussed, and it is from this analysis that the model emerged. From this same data it was also possible to analyse the language differences between the different groups of people as illustrated in the next two figures.

Even though all of these people were being asked the same or similar questions, the data clearly shows that business people and IT people actually talk about different things. Whilst business people predominantly talk about the business as indicated by the predominance of red bars in Fig. 2.3, the IT people talk about technology as indicated by the predominance of green bars in Fig. 2.4.

Each bar in these visualisations represents a different discussion where the horizontal scale indicates a relative time duration for the discussion. The bars in these figures are a small but representative sample of the whole population of discussions with business and IT people at all levels on four continents.

Business people, technology people, and academics have long been aware of the communications problem between business people and IT people. Now the science is confirming that there is a problem and the extent of it. This has perhaps always existed but has almost certainly been exacerbated by the separation of business and IT. Although this was uncovered by analysing dialogue, we know that the problem also extends to service level agreements which are often expressed in terms that cannot easily be understood by the business people for whom the services are intended.

This can only be fixed by changing the way we work.

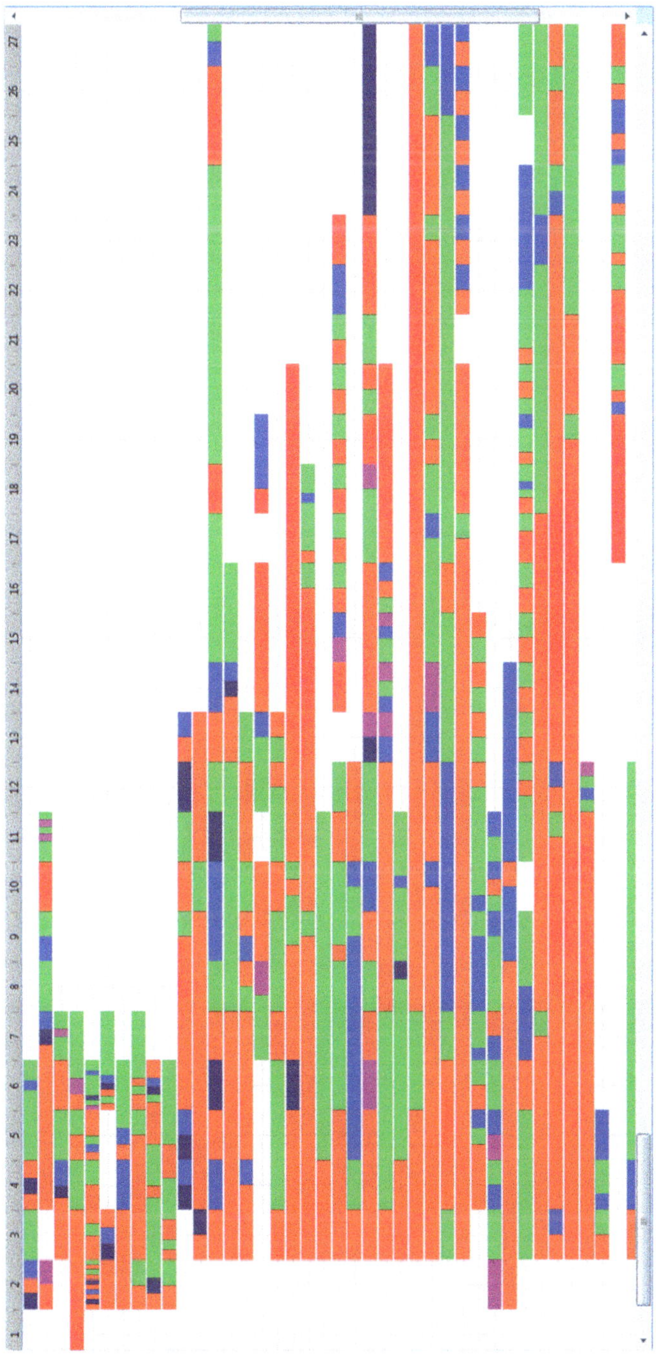

Fig. 2.3 Visualisation of discussions with business people

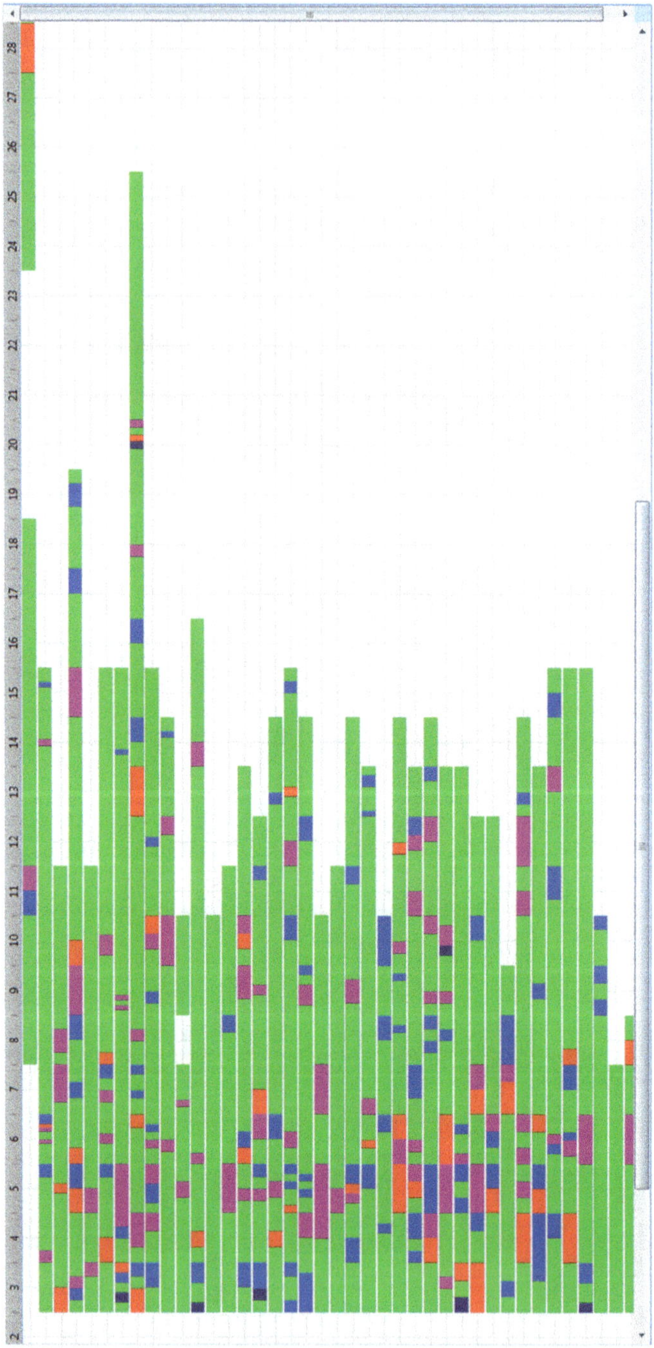

Fig. 2.4 Visualisation of discussions with technology people

Spectrum of Business Requirement

When introducing the BTRM, the logic implied a sequence of events to build what was referred to as the Causal Chain. When we are setting out to build systems in the conventional way, we like to understand this logical progression, however, our needs don't always arise in such an ordered way and are rarely explicitly linked to a Change.

An alternative interpretation of the BTRM is that business requirements can be expressed in different ways and at different times by different people. Together Business Need, the TBE, and the Service Requirement can be referred to as the 'Spectrum of Business Requirement', see Fig. 2.5. This amounts to a statement of different types of demand but rather than being created in a sequenced way and linked to a Change as described above, it may be assembled as requirements are identified. A new or changed Business Need is usually explained by the business context, strategy, and the Needs Gap and is a top-down approach. A new need can be signposted or inferred if an aspect of the TBE indicates some level of dissatisfaction combined with an explanation and is often an indication of an operational difficulty of some kind. Finally, a need may be expressed as a request for a service of some kind in the form of a Service Requirement, e.g. where perhaps some analysis is required of a business process to establish a better way of working, or there is a need

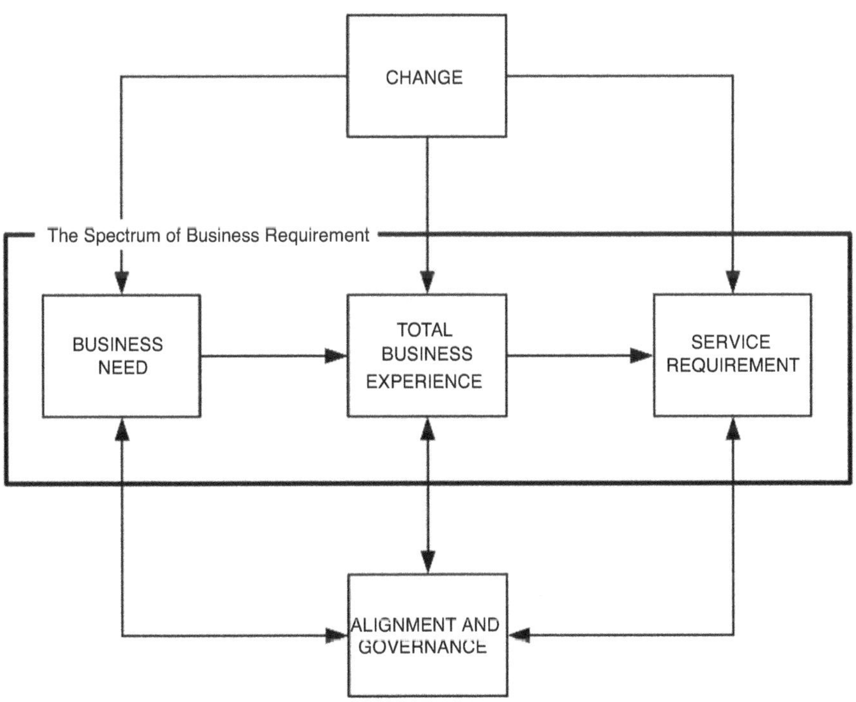

Fig. 2.5 The BTRM showing the Spectrum of Business Requirement

to build a new application, etc. By adding content not only is it possible to eventually build, perhaps through collaboration, a comprehensive Spectrum of Business Requirement but a coherent Causal Chain where the various requirement categories are linked together and linked to a Change.

As a Causal Chain the requirements are expressing the impact of Change. Causal Chains may be complete or partial. It is normal for Causal Chains to be partial initially as fragments of the Spectrum of Business Requirement are identified with a particular Change. These get completed as and when more information from more people becomes available. We will perhaps have many Causal Chains arising from a single Change, i.e. the Change will affect people across an enterprise differently depending upon their role. These may overlap such that they help to more usefully define a comprehensive Spectrum of Business Requirement, however, it may be important, as we assess the business impact of the Change, to act before we have what we consider to be a complete Causal Chain. As for the Spectrum of Business Requirement for the whole of a complex enterprise, that may never be complete.

Managing these flows of information helps business to understand when to act and how to define Alignment Tasks. Thus, Alignment and Governance becomes a key control point for the business as will be described below in the section 'BTRM as a governance framework'.

TBE: A Measure of Service and Indicator of Change

The importance of managing the total business experience, and hence the TBE, cannot be over-emphasised because it is central to the relationship between a business and its TSPs, is a comprehensive measure of service, and business decisions are often based upon this even though hitherto it has not been defined.

The TBE is the Core Category, i.e. in GTM terms it accounts for the most variation around the management and governance of enterprise technology, and reflects satisfaction levels across all aspects of business technology product and service. Many in enterprise technology are concerned about meeting expectation; they talk about 'managing expectation' and link expectation with satisfaction. There is a real sense that, under these circumstances, expectations are sometimes being kept deliberately low in order to ensure they can be met. Problematically, low expectations are associated with consistent underperformance, can inhibit strategy development, cause Needs Gaps to be understated, and limit what can be achieved. On the other hand, where business expectation is high every aspect of service can be assessed against the stakeholder's ideal requirement to produce continuous improvement.

The TBE prompts the stakeholder to consider service using a framework of nearly 80 properties across ten dimensions covering the intellectual as well as the abstract, or social and cultural, dimensions of alignment that have proven to be so elusive in the past. An algorithm is used to determine the overall TBE value—a comprehensive measure of service quality.

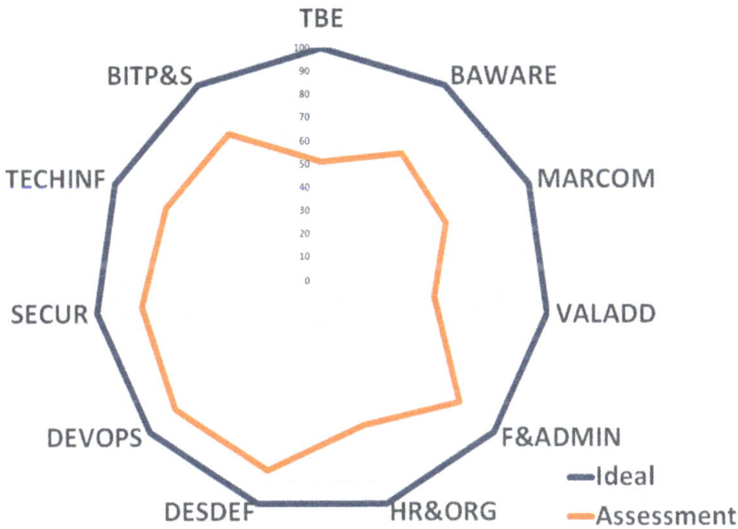

Fig. 2.6 The Total Business Experience (TBE) Radar Diagram

The process of adding value to business is now recognised to be complex and often unlocked through services where the co-creation of value is achieved by the business and technology people working together. Service means many things within the business technology context. Real service is identified as a critical aspect of the relationship between business and technology and is essential to value creation as well as value realisation and to achieving total alignment. Whilst the contribution to profits can be measured, there has previously been no measure in widespread use for managing the quality of the technology services that enable the creation of value. This measure of service quality is now provided by the TBE.

In Fig. 2.6 the outer black line indicates the ideal score and this is always shown as 100%. The red line is an individual stakeholder assessment expressed as a percentage of what the stakeholder currently considers to be the ideal for each of the ten dimensions and the score for the TBE is calculated and inserted at the top of the diagram. On the radar diagram the dimensions are sometimes identified using mnemonics as shown here but the long names corresponding to each mnemonic can be found in Fig. 2.7. If a set or subset of data is being examined, the red line will represent the mean for the set, i.e. for the business, division, or stakeholder group being viewed. This is clearly more dynamic than using absolute scores because it recognises that the scale and upper bounds may change thus reducing the score itself. This reflects reality in that groups can become dissatisfied merely because circumstances have changed for them, or something new is available, rather than because what they have is no longer doing what it was designed to do. This can happen as a result of Change and there is more about the TBE as an early indicator of Change later in this section.

Dimension Short Name	Dimension Long Name	Assessment Score	Dimension Short Name	Dimension Long Name	Assessment Score
DESDEF	Design & definition of products and services	85	BAWARE	Business awareness	65
DEVOPS	Development and operational delivery of products and services	85	MARCOM	Marketing and communications	60
SECUR	Security	80	VALADD	Sales and value add	50
TECHINF	Technical infrastructure	75	F&ADMIN	Finance and administration	80
BITP&S	Bought-in technology products and services	75	HR&ORG	People, organisation and culture	65
	CORE SERVICES TOTAL (C)	80		SERVICE MANAGEMENT TOTAL (M)	64
				TOTAL BUSINESS EXPERIENCE (TBE)	**51**

Fig. 2.7 An example of a TBE assessment calculation

Assessing the TBE against the 'ideal' provides a method of measuring service quality whilst avoiding the complications of the alternatives, e.g. by determining expectation minus perception as in the 'disconfirmation model', by using a narrow definition of service, abstract terms or constructs, and complexities arising from differing 'Likert scale' comparisons, etc. Regardless of the technology deployed or IT methods used, the business management can take a position on the adequacy of the resulting business experience and define what must be done to bring this closer to the needs of the business community. This need for a business focus also explains why assessments that are undertaken by the TSP alone are unlikely to be credible.

It's important not to confuse the TBE with those expressions of 'experience' which are often short surveys undertaken as a follow-up to a customer service call. By contrast the TBE is a measure of real service quality and is provably correct. As and when alignment tasks are completed assessments should return to their former levels. Being able to accurately assess the business experience is therefore important and the method used for calculating assessments of the business experience must be explained.

The approach and rationale is similar to that discussed by Yuval Noah Harari in Homo Deus (Harari, 2016) who describes Kahneman and Redelmeier's work on pain levels explaining that the part of the brain used for narrating doesn't aggregate experiences—it averages them. Thus, as Harari explains, experience can be a sophisticated mix of fact and myth. It is only by disaggregating experiences, gathering evidence to support the claims, and seeing others with similar experiences that we are able to understand them.

So how are business experiences disaggregated? The BTRM classifies stakeholder experience across a number of different dimensions from across the business. These individual experiences can be averaged to provide knowledge about different sets of stakeholders. Thus, whereas TSPs have measured their outputs to ensure that the Service Level Agreement (SLA) is being met, the business now has a formal and provably correct means of determining whether those services continue to meet its needs in every part of the business. Business experiences can be further disaggregated by considering the different dimensions of service as defined by the TBE.

Making sense of technology services by disaggregating the dimensions of service is now a well-defined process. The first implementation was described in my earlier work, Business Focused IT and Service Excellence (Miller, 2008). It was confirmed and refined by the research. These dimensions fall neatly into two halves. On the left-hand side of the radar diagram (Fig. 2.6), we have the five core services associated with the delivery of the technology and on the right-hand side are the five aspects which most influence service management, or what we might think of as brand including the abstract, or social and cultural, dimensions of alignment that have so far been largely ignored in management methods. The TBE is the product of the two, thus a low score on either the delivery of the technology or service management will yield a low overall TBE score. A detailed illustration of how these dimensions might be assessed and the TBE calculated is shown in Fig. 2.7. The scores for the individual dimensions are determined by calculating the arithmetic mean of the ratings given by the stakeholder for each property of that dimension (there are on average eight properties to each dimension), or by averaging the ratings from every score in the set being analysed. Most of the properties are particular to the dimension but every dimension has a budget and leadership property, even where there is no corresponding organisational unit for that dimension. For the detail of the dimensions and their properties, as well as the ontology for the rest of the BTRM, the reader should refer to the appendix.

Each dimension is scored out of 100 from the scores given to each property. It represents the extent to which that dimension meets the ideals required by that stakeholder. These are collected from individual stakeholders from across the extended enterprise, ideally including from the TSP. The TBE scores may be aggregated to represent a consensus view of a particular group, division, or the whole business. Scores are important because they draw attention to potential problem areas, however, in all cases the stakeholder scores are supplemented with a free text explanation which offers stakeholders an infinite range of possibilities for describing the problem or opportunity and thus potentially a more practical amplification of the assessment score for any property of any dimension. This is particularly valuable for service management activities where qualitative judgements are more prevalent. The improvement and developmental ideas in the textual explanations, as well as the scores, are consolidated and can be subjected to importance/performance assessments which result in new Alignment Tasks.

The average score for the core services (C) and the service management (M) are the statistical mean of the dimensions of which they are comprised. The Total Business Experience (TBE) is calculated as the product of the score for the core service and the service management and expressed as a percentage of the ideal, i.e.

$$C = (85 + 85 + 80 + 75 + 75)/5 = 80\%$$

$$M = (65 + 60 + 50 + 80 + 65)/5 = 64\%$$

$$TBE = (80 \times 64)/100 = 51\%$$

Thus, the Total Business Experience (TBE) is a function of the core service and the service management, i.e. $f(C, M)$. There are parallels here with the marketing of professional services (Kotler et al., 2002). We could argue that the TBE is a function of the price-value relationship of the core service and the brand value created by its externally facing activities. We have substituted the term 'core services' (C) for 'price-value relationship', and 'service management' (M) for the 'brand value' of the externally facing activities.

The argument to support this approach to arriving at an overall score is strong one. The empirical evidence from several commercial case studies suggests that poor service management actually reduces the impact of the core service, and vice versa, and that this formula yields the closest expression of the functional relationship between the core services, service management, and the TBE. Thus, scores of 90% for C and M would more reasonably yield a TBE of 81%, not 90. If the score of either C or M was zero, the TBE would correctly be zero. The scores for the TBE contrast markedly with those from typical customer transaction surveys or customer satisfaction surveys which are often treated as public relations exercises and where much higher scores can be expected. Alternative assessment calculation options may be used provided the formulae are clear and unambiguous, however, it must be noted that the solution described for determining service quality has been subjected to peer review and produces good results in practice.

From a business improvement perspective, the output from an assessment and discussion of the TBE is likely to yield many more improvement opportunities for bringing service to an ideal standard than traditional methods. For example, simplistically asking the question 'What do you want?' is always unsatisfactory because it happens at a time that suits the TSP when the business person may not know in detail what is required, it is typically asked in one to one or one to few contexts, and there is little opportunity for collaboration. Using the BTRM instead as a framework for collecting the data is a great way to improve agile methods; how often are we in an agile workshop where the person thought to have the answers is not in the room? Perhaps because of the comprehensive nature of the TBE, and because scoring puts the onus on stakeholders to explain their claims, subjectivity is minimised such that there has never been any dispute about the results of the analysis amongst stakeholders. This has enabled the model to be frequently used in workshop sessions following minimal explanation with good results making this approach a useful adjunct to traditional methods where a meaningful measurement of service is needed. This would perhaps be used to baseline a service before developmental work was undertaken, to establish the strengths and weaknesses of a service, or simply to explore what needs to be done.

If the TBE is a measure of service, then it has the potential to change the way in which commercial contractual arrangements are managed. There is some dissatisfaction with service contracts, or Service Level Agreements (SLAs), in that they are often couched in technical terms that are not always understood by business or cannot be measured or verified by the business whereas the TBE makes sense to business people.

The importance of managing the total business experience, and hence the TBE, cannot be over-emphasised because it is also an early indicator of Change, even where a Change is not yet explicitly stated. How does this work and how can it help? When scores reduce, the TBE is telling us that some aspect of the service is no longer meeting what is ideally needed—either to meet today's needs or those in the near future. This could have arisen as a consequence of a Change having occurred, an expected Change, and/or some newly identified Business Need (and this will be clear from the explanation in the assessment).

The TBE is sensitive to everything that is happening externally and internally that might lead to services not meeting the ideal required by business and so a revised assessment of the TBE can precede any formal declaration that Change may be taking place. For example, if people in the business become aware of some new technology that is available to their competitors, then they will eventually begin to register dissatisfaction with their own systems; when someone sees a market opportunity and the current technical infrastructure precludes the business from entering that market the stakeholder's next assessment of the TBE will be downwards to reflect that. Both would be likely even if the Change had not explicitly been entered.

Case studies show that Change was explicitly reported to the TSP in less than 2% of cases, and certainly not at an early stage. May be in the vast majority of organisations the IT function does not require that information, or does not respond until the business can provide a full specification of their requirements. In contrast to this there are many examples where a Change could be implied as a result of assessments received.

By continuously assessing or monitoring the TBE it is possible to detect when Change is being anticipated, detect when it may have occurred, understand the business impact of Change, and ensure an agile response in order to maintain a managed business experience—even when the Change itself has not been explicitly reported. This characteristic has the potential to provide a safeguard for the business where self-adaptive and self-managing systems are deployed or where other complexity within the business and technology landscape may hide what is taking place.

When Change is explicitly reported early, or when using the TBE as an early indicator of Change, the TSP and business people have an opportunity to collaborate sooner, the business gets a solution sooner, and the whole enterprise becomes more agile.

Blind Spots Causing Failure

It has been possible to understand more about today's management failings by looking at how current processes typically map to the BTRM.

From a business perspective, from a Business Need business people will identify a problem area and work with the TSP to build a specification. The process can take a long time and things can change. The process will probably start late because the business has been conditioned not to approach the TSP until it knows what it wants, and a limited number of people are involved in the process so the specification will be incomplete anyway. Initiatives are rarely linked to external events which would provide context and so business investment decisions are often difficult. Requirements in the same problem area can be aggregated so inflating resource requirements, timescales, and cost. This dominant business perspective is highlighted in Fig. 2.8.

From the service provider's perspective, they will see that a new technology has emerged and has become viable and it considers whether or not it will benefit the business. It may have already come to the attention of the business if it was known to be being used by competitors but otherwise it would be a service provider initiative. The TSP may not yet have been told of the current Business Needs or have a complete picture of levels of satisfaction across the enterprise of existing systems and so exercising judgement as to which technologies will present business

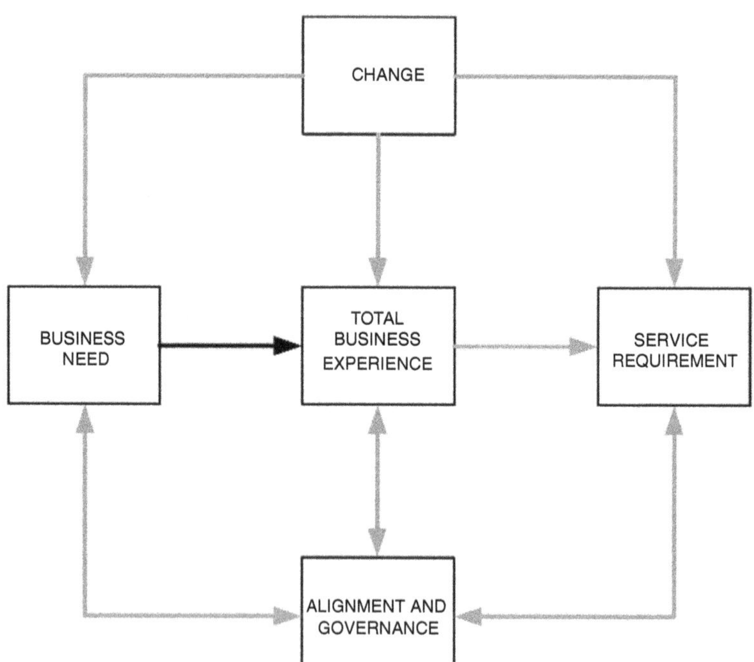

Fig. 2.8 The dominant business perspective of the BTRM

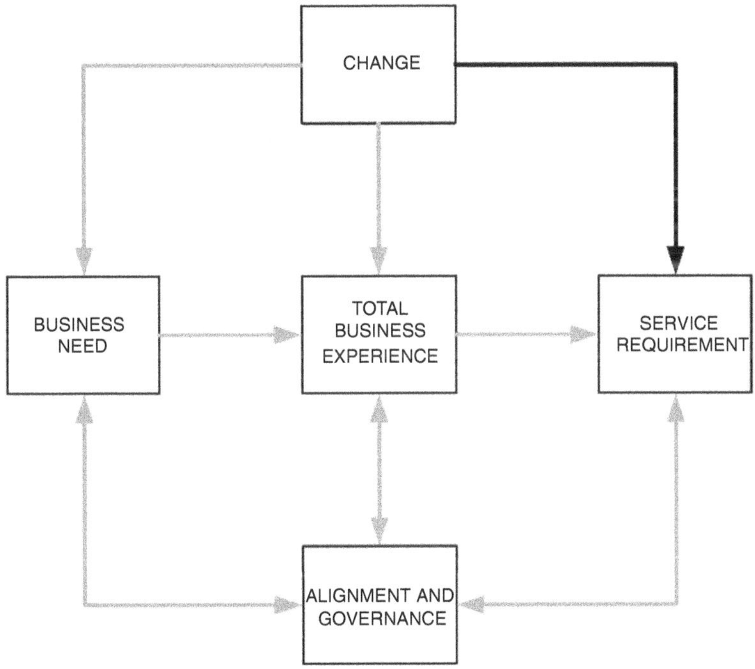

Fig. 2.9 The dominant service provider's perspective of the BTRM

opportunity and how and where these should be deployed is difficult. Often, those who shout the loudest get the investment and this is not always appropriate but by then resources will have been wasted. This dominant service provider's perspective is illustrated in Fig. 2.9.

Even in combination the combined perspectives provide an inadequate view of the business and at best this can make IT management look weak. At worst IT management, or its methods, can be seen to be contributing to failure. So far, management hasn't dealt well with the increasingly complex, ambiguous, and dynamic real-world relationship that now exists between business and technology, and between TSPs and the business stakeholders. We need to change the way we work.

The BTRM as a Governance Framework

Alignment and Governance regularly feature in the top 10 issues for the CIO. The governance of IT and technology generally is also a business issue but neither the CIO nor the business can carry out their responsibilities in this regard, if, as explained in Chap. 1, both alignment and governance are inadequately defined. Under current management regimes, alignment is seen to be a helpful planning activity but for some at least its use of static constructs makes this less useful.

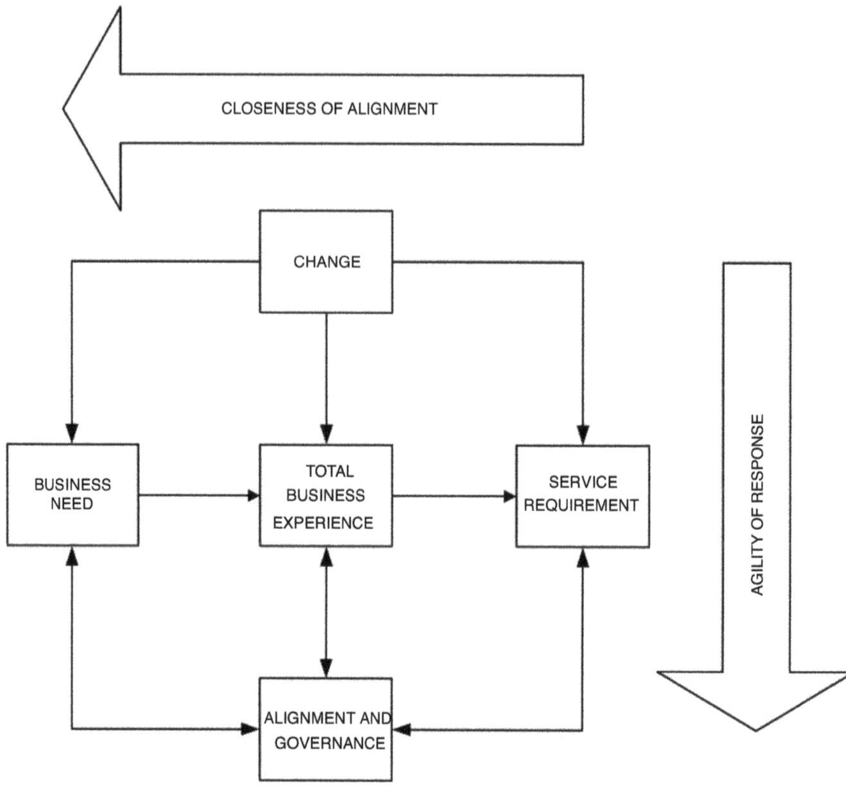

Fig. 2.10 The BTRM showing the themes of Closeness and Agility

Governance most typically is a post implementation or an infrequent audit activity perhaps carried out to establish whether some IT method has been followed. For the reasons given in Chap. 1, trying to judge good governance under these circumstances is probably, in most cases, not going to help the business. So how does this change with the introduction of the BTRM? If we consider the BTRM as an over-arching management framework, then Alignment and Governance can be based upon the concepts of 'Change' and 'Response to Change' and two themes are apparent, see Fig. 2.10.

The first can be considered to be the current status of the Spectrum and Business Requirement and this theme represents the difference between what exists and what is needed and is termed 'Closeness of Alignment'. Total alignment occurs when the Spectrum of Business Requirement precisely meets the needs of all of the business stakeholders. Closeness of Alignment is suggested as an expression of the extent of alignment at a point in time, which thus becomes a key measure addressing the concern in the literature that alignment is an elusive concept. The second theme is one of performance and represents the time between a Change being identified or notified and an Alignment Task being successfully deployed which addresses the

Spectrum of Business Requirement that is consequential to the Change and is termed the 'Agility of Response'. This addresses the confusion surrounding the term 'agility' which most often today is applied to the development time.

Thus, regardless of the IT methods used the BTRM tells us to assess that what is delivered is what is needed and that it gets delivered in a timely manner. For the business this is a provably correct process and the results can be measured, not annually but continuously. Hence if we change the way we work to use the BTRM as a framework, Alignment and Governance becomes a key control point for the business actively managing the TBE through the creation of Alignment Tasks according to what stakeholders are telling us is what is needed. Thus, we have a mechanism for the early provision of solutions and services, and a means to monitor their timeliness and effectiveness. Alignment and governance are now defined and the measures of success are the same and obvious to all who are involved.

Changing the Storyline

The BTRM offers the prospect of a new management philosophy that enables us to change the way we work. Instead of offering this as a methodology as in the past, its characteristics enable this management philosophy to be encapsulated within a management system. The working name for this conceptual system is 'Active Management and Governance'. The system provides an improved basis for the practical assessment, alignment, management and governance of business technology, or any complex service in any domain where technology is ubiquitous. It avoids the problems of different groups talking about different things, of maybe not involving the right people, of perhaps aggregating to create a mammoth project rather than disaggregating to deliver the important needs early. It allows us to change the way we work and in doing so we also change the storyline.

The storyline today, using current methods, will be familiar to most people, most notably to CIOs and business management. It was summarised in the opening chapter as one of the unintended consequences of today's management methods. Set out more formally it is as follows:

I. Services are delivered to the business to meet an implicit or explicit need.
II. When enough information is available fixed specifications determine what system or service changes are required.
III. After a system or service is implemented, the business will expect what is being delivered to be closer to its needs—but often the process takes too long or other objectives are missed.
IV. Changes external to the business, e.g. in the market generally and in the world of technology, influence the Business Need potentially creating a need for further realignment—sometimes before the current realignment is complete.
V. Technology Service Providers are always playing catch-up not knowing where the next requirement is going to emerge.

The contrasting storyline when using Active Management and Governance to achieve a managed business experience is as follows:

1. Services are delivered to the business to meet an implicit or explicit need.
2. The services are continuously assessed by the business stakeholders and others in order to determine the Total Business Experience.
3. Everyone can contribute and gaps between ideal needs and the assessments indicating a lack of alignment will be investigated to identify where the need or impact has occurred, what Change may have occurred to give rise to this, and what else may be affected.
4. Long-term plans will be reviewed and short-term action instigated where necessary by adjusting the service provision to create value and achieve realignment.
5. Alignment and Governance is a dynamic key control point for the business for JIT decision-making actively managed through the creation of Alignment Tasks.
6. The Closeness of Alignment and the Agility of Response are key to both stakeholder satisfaction and service performance and are actively managed by the business and the Technology Service Providers.

Chapter 3, Active Management and Governance, describes the basic functionality of a conceptual solution based upon the BTRM and the developmental work that followed the discovery. It is illustrative of the manner in which this revised storyline can be implemented in the context of a future complex organisation.

References

Brooks, T. (2000). *Star wars episode 1 the phantom menace*. Arrow Books.
Harari, Y. N. (2016). *Homo Deus: A brief history of tomorrow*. Harper.
Kotler, P., Hayes, T., & Bloom, P. (2002). *Marketing professional services* (2nd ed.). Prentice Hall.
Miller, D. (2008). *Business-focused IT and service excellence* (2nd ed.). British Computer Society.

Active Management and Governance 3

The previous chapters described what is simultaneously an IT management problem and a business problem and introduced the BTRM as a means of considering an alternative management philosophy that provides an opportunity for us to improve management practice by changing the way we work. This chapter describes the basic functionality of Active Management and Governance, a conceptual solution for the management of business technology based upon the BTRM and the developmental thinking that inspired. It assumes it is used by all stakeholders across the extended enterprise so allowing us all to change the way we work.

In discussing the functionality in any illustrations that follow, we have to make some assumptions about the purpose of this implementation of the Active Management and Governance philosophy. It could either be introduced as an over-arching framework to existing methods (though they will inevitably have to adapt), as a front end to new methods, as a control, or as a control on total automation. As we don't know in detail what new methods or automation will look like, the simplest assumption to make that everyone will understand is if it is to be used as an over-arching framework to existing methods before decisions are made, through the experience of using it, as to how we wish to adapt them.

Section one 'Roles within a changing enterprise' discusses the nature of today's enterprise, its stakeholders, their organisational context, and the roles they play in this new management system, section two 'Setting enterprise policy' explains the freedoms that can be assigned to stakeholders, and section three 'Knowing the stakeholder' explains more about the stakeholders organisational context and how this can be managed in an easy to use manner, section four 'The Active Management and Governance dashboard' describes the multiplex display used for input and analysis, section five 'Capturing the management entry' looks particularly at how Change and Business Needs are entered, section six 'Capturing assessments of the Total Business Experience' describes the basic mechanics of entering an assessment, and section seven 'Capturing the Service Requirements', and section eight 'Establishing and recording alignment initiatives' complete the introduction to Active Management and Governance. Section nine 'More about value-in-use'

D. Miller, *Managing Agile Business Technology*, Management for Professionals,
https://doi.org/10.1007/978-3-030-90598-9_3

explains the importance of continuous assessment and introduces Chap. 4 which opens up new possibilities and the prospect of moving from a serial process to a real-world context of multiple parallel processes.

Roles Within a Changing Enterprise

In the past we would have been very concerned to observe the internal divisions and departments of the organisation which might have more to do with the history of the business than with what it does and how it operates today. In that organisation would be directors and executives, managers, process owners, process operatives, system owners, data owners, analysts, and other specialists. Today we manage process across organisational boundaries. Many enterprises will be dependent upon many external communities. There may be multiple external suppliers and even multiple TSPs, customer businesses, and consumers of the end product or service. Whatever their role in the extended enterprise, together these are all stakeholders operating in different groups but perhaps with different but related interests. In addition to this there is a tendency for increased acquisition and merger activity which takes groups and moves them to new enterprises. This evolving nature of the enterprise changes the way in which we understand process and manage systems.

We saw in Chap. 1 that there is a need for the management and governance of technology to involve the business more, to address a broader spectrum of business needs, to connect with business more effectively, and at more levels, and for this to be a continuous and even, where relevant, a more immediate process. Thus, just as any business desires an open and collaborative culture within the workforce in order to get information flowing upwards to help improve business, in Active Management and Governance we would wish to actively encourage this data flow across the different communities. With the right information flows and the BTRM as the management framework a business can manage and exercise governance in an agile manner.

In many large businesses there is a board of directors and below that the Executive which is the group of senior people responsible for operational activities. Historically, the group of people responsible for setting the direction of investment in business technology based on strategy and information flows and who prepare the recommendations to the Executive for investment in technology tends to be drawn from amongst the technologists and they include architects and other planners.

In Active Management and Governance, this is the work of the Alignment and Governance category of the BTRM. As was described in the last chapter, Alignment and Governance becomes a key control point for the business. As such, with suitable tech-savvy business people at the Executive level this should be a business responsibility. If in the shorter term the Executive does not step up to this task, then there is a requirement for it to work very closely with the strategists and the latter must be tech-savvy business-focused people. To maintain this distinction from what has happened in the past, in this publication, this tech-savvy business-focused team that is undertaking the work of the executive is referred to as the Gold Team. If kept

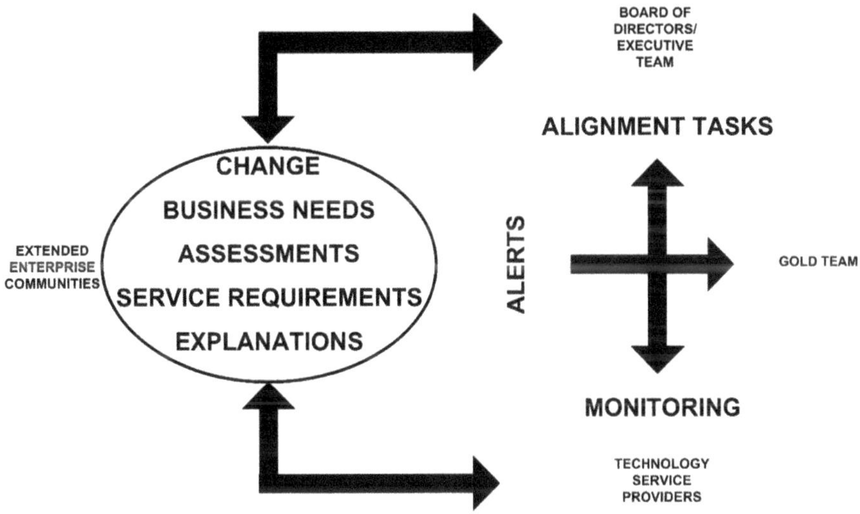

Fig. 3.1 Overview of information flows

separate initially then members of the Executive could be co-opted to the Gold Team to speed up the transition to its ultimate role. The Executive/Gold Team will be accountable to the Board of Directors for making investment decisions by initiating alignment tasks to ensure the technology meets the needs of the business—and will not just monitor the results but be able to react to them. Alerts may be set to bring out-of-tolerance conditions to the attention of the Gold Team and the Executive. With better information, solutions can be provided more quickly and alignment can be fine-tuned and monitored, see Fig. 3.1.

Greater agility demands streamlined methods and at the heart of every business is how the organisation sees itself and its extended enterprise. To make sense of the dynamics of the BTRM there is a need for a mechanism for understanding the logical structure of the whole of a large and complex extended enterprise without the, sometimes, unnecessary detail of the company organisation chart. Everyone in business is a stakeholder of some kind and these can be separated into Key Stakeholder Groups based on the principal business activity. These may not correspond precisely to the organisation chart. Each Key Stakeholder Group (KSG) within the extended enterprise community will have access to flows concerning their own activities whilst the most senior people in the business will need access to the constant stream of information from potentially all stakeholders across the whole of the extended enterprise. The origin of the information flow, i.e. the KSG of the stakeholder making the entry, provides important context.

The KSG graphic has been devised specifically for Active Management and Governance. It is a simplification of a true organisation chart that enables more information to be held in a structured, efficient manner, which is more manageable and more visually accessible and has the advantage that it can be applied to any community. The internal organisation chart also sometimes has more to do with history, politics, and people issues than it does with the key activities of the business. The lines of an organisation chart can take on more significance than the boxes whereas the BTRM philosophy is more concerned with what the stakeholders do and how the technology supports that. Because we are not concerned with lines and don't need to know them, a KSG can be inside or outside of the organisation and defined as we please.

In the simple example shown in Fig. 3.2, on the left is a conventional organisation chart for a single division of the imaginary Dave Corporation alongside a KSG chart on the right for the same division. In this example the KSGs match the organisational departments but that need not be the case. As we add more divisions to the KSG Chart we add more columns and for visual effect we use a different infill colour. Depending on the nature of the Dave Corporation we can arrange the KSGs to emphasise the importance, e.g. they can be arranged in the form of a simple value network with primary activities first followed by support activities, or as above and below the line activities. More complex examples will be used later to illustrate the power of this representation. The simple example in Fig. 3.2 only shows one internal division (the Dave Servicing Division) in order to illustrate the concept but later we will show the KSG chart for the whole of the extended enterprise of the Dave Corporation.

Maintaining detailed organisational charts for large-scale enterprises is difficult. KSGs, however, are logical activity groups and new ones can simply be added as appropriate and assigned to the division to which they belong by assigning the appropriate infill colour. Super groups can also be created. These sit at the same level as the division and they are used to group together certain external KSGs for the convenience of analysis, e.g. business suppliers, key business customers, etc. The row and column in which KSGs appear are not significant though for convenience there would be an attempt to group KSGs belonging to the same division or super group together. The colour coding technique, whilst removing the complexities of the organisational chart, still enables the desired group, divisional, or companywide perspectives to be easily identified and selected for analysis. External KSGs (including any external TSPs) are defined according to needs and the appropriate super group can be used to compare the experiences of different types of stakeholders across the extended enterprise, e.g. an internal division with key business customers, or an internal division with an external TSP.

Typically, businesses will want the problem definition and requirement to be shaped from the first moment that a Change is reported or the first indication of a Needs Gap being received regardless of whether there is a complete specification at that time, hence the importance of early information to build the Spectrum of Business Requirement. Senior people may want to encourage this by entering their thoughts about Change early to prompt responses. The Business Need, a modified assessment of the TBE, and the Service Requirement will be defined, perhaps from multiple perspectives across the extended enterprise. These form the basis of an initial 'go and see' list (what Toyota calls *genchigenbutsu*), and subsequently of a project brief, and they will be assessed for dependencies, prioritised, and will compete for resource in the normal way. Not only does the response start sooner

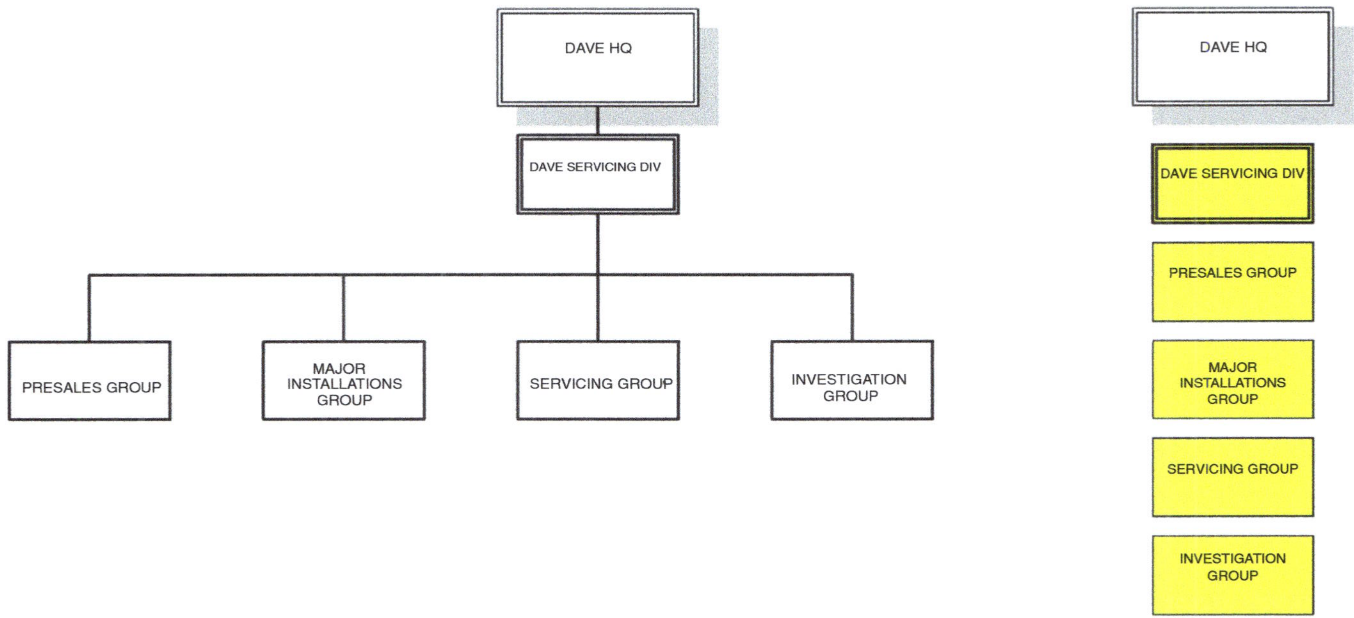

Fig. 3.2 Organisation Chart and Key Stakeholder Group Comparison

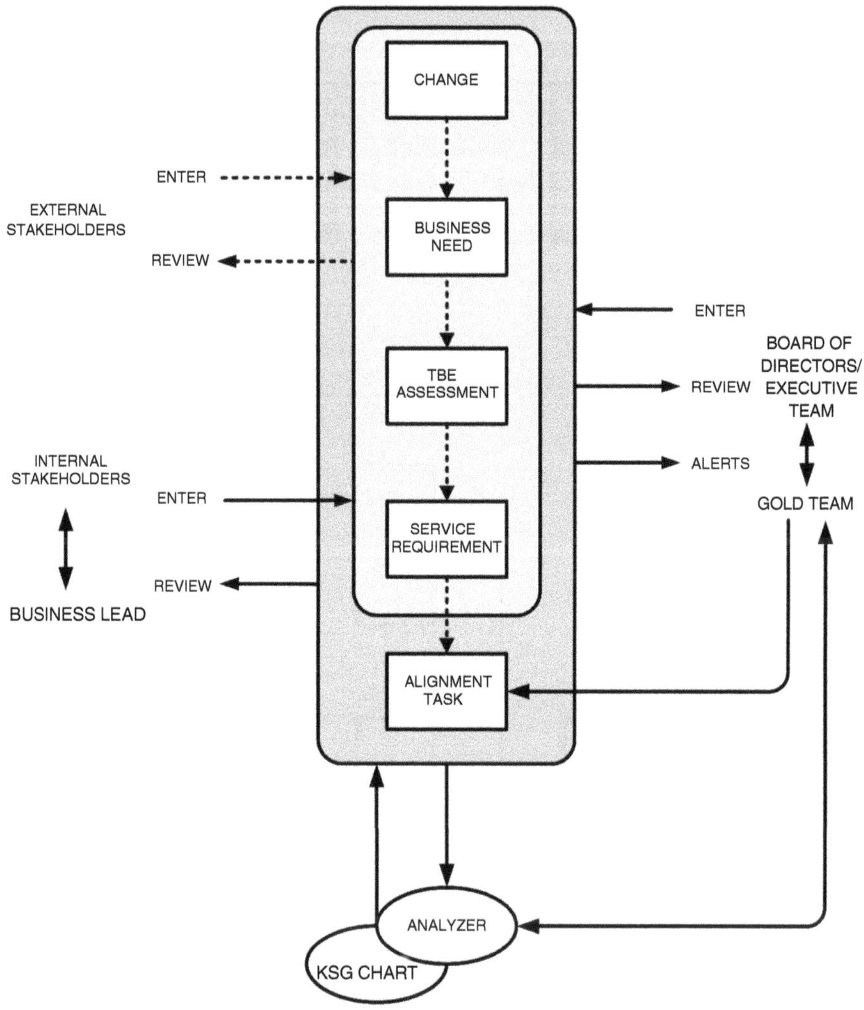

Fig. 3.3 Simplified use-case diagram

but the situation is constantly kept up to date with the latest intelligence from across the whole of the extended enterprise and plans can be modified dynamically to reflect the changing circumstances. This addresses the conundrum of how agile businesses can work with business technology and TSPs. The key control point for the business becomes the Alignment and Governance processes controlled by the Executive/Gold Team. Representatives of the Executive/Gold Team will interact with the other key stakeholders. The various types of stakeholders and their involvement are illustrated in a simplified use-case diagram in Fig. 3.3.

The figure shows five generic classes of stakeholder. The business lead is a senior manager who sets direction for a part of the business appearing in the KSG graphic, i.e. is responsible for a Key Stakeholder Group or higher grouping of these, e.g. a division. An internal stakeholder will be a member of a named KSG within the enterprise, this could either be a business stakeholder or a stakeholder within the internal TSP. An external stakeholder will be a member of an external entity again defined by a named KSG, typically either a business customer, business supplier, or external TSP. Other stakeholders are members of the Board of Directors, their Executive, and the Gold Team members themselves.

As indicated by the simplified use-case diagram in Fig. 3.3 by default the business lead and the internal stakeholders (including those within the main or only TSP) may create any kind of entry except for Alignment Tasks and access all entries. It would not be normal for external stakeholders in the business supply chain or for key business customers to see each other's entries, nor would it be normal for them to see the Alignment Tasks. Having suggested what is normal, who can enter what and who sees what is a policy decision for each enterprise (see the section: 'Setting Enterprise Policy', below).

Directors and the executive have the ability to create data, e.g. to record Change, and review anything entered by others. They would be particularly interested in aggregated data and comparisons. The Gold Team on behalf of the board/executive would also have the ability to monitor all activity, create alignment tasks, and maintain the KSG Chart. To help them in this process the Gold Team would also have comprehensive analysis tools to help them to navigate through the data (here called the 'Analyzer', which is explained further in Chap. 4, Analysing the Real-World Landscape). Additionally, alerts can be arranged whenever assessments, or whenever aggregations of assessments, exceed a desired threshold. Although optional for most stakeholders these alerts would be of interest to members of the board/the Executive and would be used by the Gold Team to ensure that unsatisfactory experiences in any community are not overlooked.

Once projects are initiated the 'normal' design and delivery methods preferred by the enterprise and service provider can be used but instead of development taking place in a vacuum, the context is maintained and the Gold Team and developers are able to continue to assess the relevance of projects at all stages prior to implementation. Internal Stakeholders can also review the progress of Alignment Tasks that are relevant to their Spectrum of Business Requirement. Post implementation, the Executive/Gold Team and the Board of Directors would expect to see improved TBE assessments from the stakeholders benefiting from those Alignment Tasks.

If our first implementation of Active Management and Governance is an overarching framework for our existing methods, we can expect two areas to change quite early, namely the activities associated with design, development, and operational delivery, and those associated with financial approvals and budgeting.

Technology development and operational delivery methods will inevitably evolve rapidly under this management regime with smaller projects and further automation being introduced into processes. Even total automation at some point may be possible, however there will still be a need to identify what Change is taking place, for sensing the impact of that Change, for monitoring the response, and placing controls over the outcome, in particular to detect any unintended consequences of automation. Active Management and Governance will continue to be relevant ensuring that the dynamics of the big picture are always visible, and that action is timely and always aligned with what is needed by the business regardless of how this is achieved.

Budgeting was always a part of the alignment and governance process. It would probably have been undertaken annually and the review/planning cycle was almost secondary to that of setting the budget. Often, the budget would soon be out of date as plans would have to be revised to meet with requests for investment. To ensure that Alignment and Governance can now be a continuous management process involving people at all levels throughout the extended enterprise ensuring that technology products and services continuously and effectively meet the needs of the business this must also be supported by flexible budgeting and fast financial approval if investment is to be timely, responses agile, and alignment maintained.

Setting Enterprise Policy

There are a number of general principles which can apply to the working of any system. Here consideration is given to where the business can set policy according to how it wishes people across the extended enterprise to behave.

There is a general principle that assessments are unique to a named identifiable stakeholder, in a job role, within a KSG, at a point in time. That person in that same context will provide many assessments over time and potentially more in the context of other positions which that person may at some future stage hold. Each assessment helps to build a profile of the Spectrum of Business Requirement for the KSG in question. Optionally stakeholders in key business customer, business supplier, and internal/external TSP organisations may participate on the same basis. For both internal and external stakeholders, it is necessary to set policies relating to who can do and see what.

The default situation allows any stakeholder within the business (including internal administration functions and the internal/main TSP) to see the entries of all others within that business. This default position allows insight to be shared amongst the largest community, encourages collaboration, and allows consensus to develop faster. The enterprise may, however, set alternative policies that limit what a stakeholder within the business is able to see, e.g. may only see entries from the business lead of the stakeholders own KSG; may see all entries in own line of management (see note below) to a defined level, i.e. all in own KSG, all in own division, all in own division plus HQ level; or entries in KSGs with which the stakeholder's KSG has a business relationship (inside or outside the stakeholders own line of management). Policies could be applied in combination, e.g. restricted to own line of management and KSGs with which the stakeholder's KSG has an internal business relationship. Note that the 'line of management' is defined by the way groups are clustered in the KSG graphic (the same line of management using the same infill colour as described in the previous section) and perhaps by nominating a business lead who, as was suggested earlier, may be given some responsibility for the creation of Change entries. The entries by external stakeholders are more sensitive and a different default and options apply.

External stakeholders from the business supply chain or a business customer organisation cannot by default see the entries by other external stakeholders from the same or any other third-party company including other business suppliers, customers of the business, or third-party TSPs. Assessments entered by external business suppliers and external business customers will be retained as separate assessments but they are aggregated for use within the business in order to maintain an average score for all key customers, all key suppliers, or all external TSPs respectively (just as for a division of the main business), thus allowing averages to be compared. By default, internal business stakeholders are able to see the assessments entered by external stakeholders. Alternatively, a business may restrict access to those external entries to those KSGs where there is a business relationship between the groups. Optionally, where there is a very close business relationship, assessments from external KSGs may be included in the overall assessments of the 'owning' business unit.

The question with external TSPs, e.g. an outsourced TSP, is the extent to which they will be given visibility across the enterprise and the extent to which the enterprise has visibility across the external TSPs. Stakeholders within an external TSP are by default restricted to seeing only those entries made by internal groups to which they provide services. Likewise, it would be the default situation that internal stakeholders will only see entries from external TSPs that provide services to their KSG. External TSP personnel will have access to the entries of colleagues within their own organisation to allow full collaboration. Other variations are possible if this is not appropriate, e.g. the external TSP may be under contract as a third-party supplier providing services to the internal TSP or only TSP. Under these circumstances the third-party supplier stakeholders may have no access to business stakeholder entries.

Within the constraints defined above, stakeholders are by default able to add entries for Change, a Business Need, a TBE assessment, or a Service Requirement and these may be complete or partial, and may form a complete Causal Chain or not, i.e. the entries will be based on the information available at the time and information will not be withheld until it is considered to be complete.

At some point the Gold Team will consider that, because sufficient is known about the Spectrum of Business Requirement, or, because from what is known the impact of Change is sufficiently significant, an Alignment Task should be initiated. Only members of the Gold Team are able to create and enter Alignment Tasks. The enterprise may set a policy which restricts what entries a stakeholder is authorised to make. In particular, organisations may wish to restrict the entry of a Change or aspects of Business Need to senior managers on the basis that they direct strategy but the counter argument is that any stakeholder can have privileged insights into the market, competitor activity, and what might be done. Not having the ability to share this information could limit collaboration and inhibit opportunity and so the default policy is to allow stakeholders to make any entry except for the Alignment Tasks.

In the real world, differences will emerge between HQ strategy and insight from stakeholders and this should be transparent within the business so that discussion can take place and the appropriate action taken: Active Management and Governance can handle this very efficiently. If, however, there is a situation where senior managers believe that this openness may not be appropriate, e.g. during a major restructuring, during mergers and acquisitions, etc., it may be that the policies are changed until the situation has become public knowledge.

The enterprise may set a policy that controls the manner in which an assessment is created. This determines the manner in which a stakeholder is prompted, i.e. what the stakeholder sees when making a TBE assessment. The default is for the stakeholder to be prompted by their last assessment in the same context, i.e. same KSG and role, and to enter only the differences from that last assessment. This approach has been chosen as the default as it is the most intuitive and most efficient. Other policy options are that the stakeholder is prompted by using an assessment of their choice created by another stakeholder in the same KSG using the argument that this will encourage collaboration and allow consensus to develop faster. Another is that stakeholder assessments are always entered blind, i.e. without any prompting so not influenced by the opinions of self or others. If stakeholder assessments are always entered without access to any previously entered assessments, including the stakeholder's own previous assessments, as many reasons as possible for the score must be provided. Blind entry is not recommended for internal stakeholders but it could be suited to external stakeholder assessments where entries may be less frequent, perhaps have less content, and avoid undue influence.

All values for the TBE assessment entries default at the start to 'Close to Ideal Alignment', i.e. 100%, and stakeholders will revise this downwards to reflect their own experiences. Note that anything exceeding the ideal would be deemed unnecessary by the stakeholder and would result in a reduced assessment so the assessment never exceeds 100%. The organisation may also set a threshold below which an alert would be raised to bring the entry immediately to the attention of the KSG, Division, and Gold Team. Thus, the ideal score is expected to be 100%, assessments below this will require an explanation, and assessments below the policy threshold will raise an alert. By default, the alert threshold when set is the same for all stakeholder groups. It is likely to start off low and be increased as service levels improve. The threshold may also be varied by Division (or external business grouping) or even by KSG on the basis that different groups may turn out to have different tolerance levels or the organisation may wish to set external business customer and business supplier alert thresholds higher than internal ones.

Knowing the Stakeholder

To maintain the integrity of the system, it's important that every entry is attributable to a real person, i.e. the 'owner', including entries by Gold Team members. If we don't know who is notifying a Change, claiming a new Business Need, making a TBE Assessment, requesting a Service Requirement or creating an Alignment Task, then we have no accountability, no context and no means of contacting the stakeholder when responding or requesting more information. Furthermore, by avoiding anonymity we avoid many of the pitfalls associated with social media, i.e. false claims, abuse, trolling, etc. We also need to balance this against clarity and simplicity especially when dealing with large, complex, or extended enterprises.

To register as a stakeholder, each user or potential user must provide their given name used by the employer to identify that person (at least the first and family name unless this is insufficient as a unique identifier), their business unit (the department name as referenced by the organisation chart, the name of the key supplier company or the name of the key business customer—it must be clear which), or the name of the TSP and department, and job title. Also required are the characteristics of the job from each of the following: (1) workplace worker or mobile worker or home worker; (2) full time or part time or contractor or temp; (3) permanent role or special assignment. Finally, the group to which the organisational data belongs is selected from the Key Stakeholder Group (KSG) Chart. These details need to only be entered once unless any of this information changes. Note that once a stakeholder is registered, every entry provides the option of entering the specific job role or activity that was being undertaken when the requirement for the entry was identified if this provides further useful context.

Once the user has been registered as a stakeholder, entries and other activity will be signed by name, KSG, job title, and date. If the stakeholder is only registered in one KSG and only has one current job title, it is only necessary to enter the name and the system will complete the rest of the signature. If the stakeholder is being asked to undertake some kind of special activity but retains the job title, role can be added at any time to differentiate these entries. When signed in, there will be a visible indicator on the KSG graphic highlighting the KSG of the stakeholder signed in to the system as visible confirmation of the stakeholder's location within the enterprise.

The KSG graphic was introduced earlier and is shown here in detail in Fig. 3.4. It allows us to represent each logical group of activities and relate these to a division of the business without connecting lines. It's easy to amend and easy to locate an activity. In addition, we can add important external groups whose activities are often considered critical in some way to our business.

Visually the external groups can either be incorporated inline as represented in a value network concept as shown in the default KSG graphic referred to above, or where there is a more hands-off relationship with external groups the KSG might be redrawn with the external parts of the extended enterprise, separated and shown below the line as in Fig. 3.5, below. Although visually the two figures are different, they contain precisely the same information except for the fact that in Fig. 3.5 the Dave Technology Services Division is now an external organisation called Technology Services Plc.

Normally the KSG is the name of the KSG to which the stakeholder is permanently assigned but exceptionally this could be a KSG created for a special project. A special project might be assigned its own KSG but this is not always desirable and there are alternative ways of achieving this distinction without being too specific.

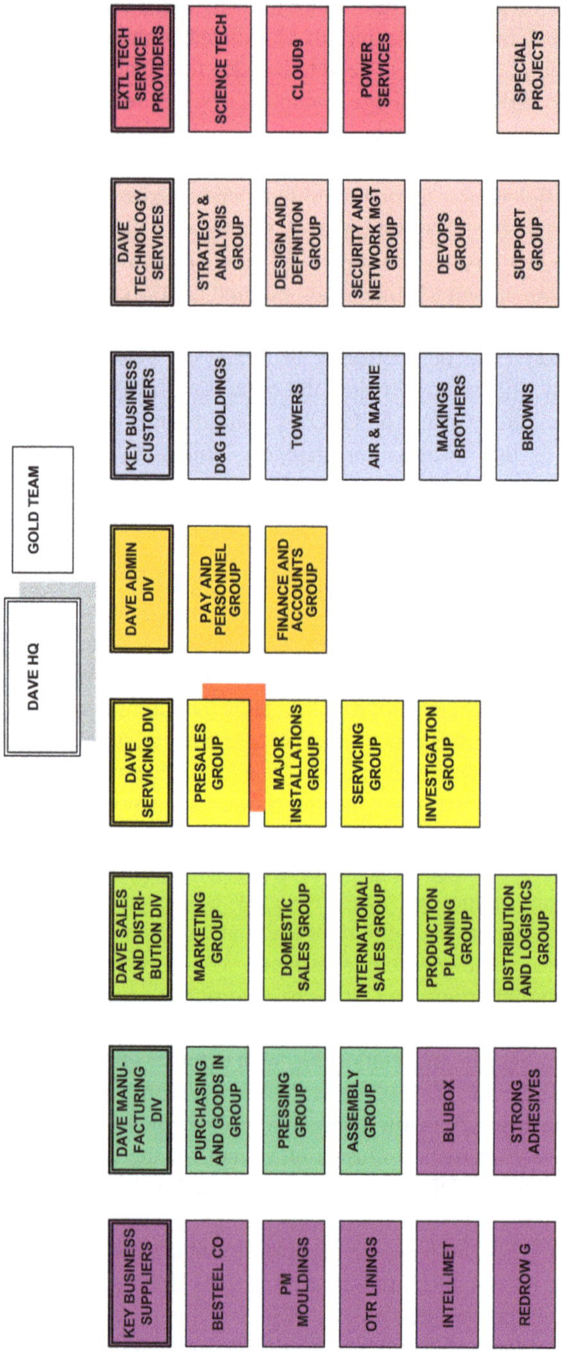

Fig. 3.4 Key Stakeholder Groups—value network format

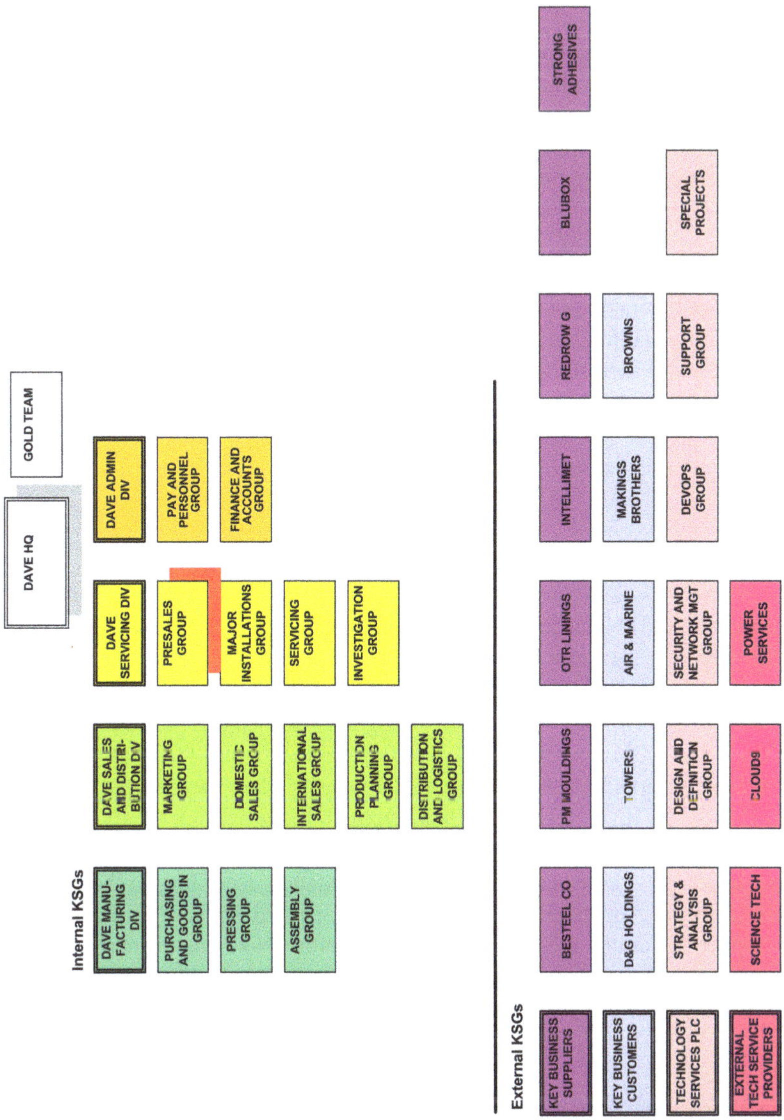

Fig. 3.5 Key Stakeholder Groups—internal/external format

The job title is normally the permanent job title but could instead be an alternative job title within a special project. Special projects come and go and management judgement will always have to be exercised when deciding whether to create a KSG for this purpose or not, and if so, when. The simple rules of thumb regarding special projects are as follows: if it should not yet be exposed to widespread scrutiny, if it is likely to be short-lived, or if it is wholly contained within an existing KSG, then creating a new KSG for the special project is probably not desirable; in these cases, 'Role' could be used to explain the context.

Roles are not defined to the system and are an optional descriptive element of the entry. The stakeholder's job role, if used, could further explain the stakeholder's specific work activity at the time of the entry to provide more information about the context of the entry, i.e. this role within the KSG and job title given that gave rise to the problem or issue being raised. Role could also describe an activity being undertaken on a temporary basis, e.g. more akin to a special project but where it was deemed not desirable to create a new KSG.

All of this information that identifies a stakeholder is shown in the heading at the top of the data entry screen, i.e. the management and governance dashboard. All of this header information, except for Role, having been entered once would be used to prefix every entry for that stakeholder along with the date and time stamp. This header information is defined in the Appendix.

The Active Management and Governance Dashboard

The management dashboard for data entry combines all of the summary data on the one screen whilst at the same time supporting data entry, navigation, and analysis. There are four graphics: these are the KSG chart with the stakeholder's KSG highlighted (top left); the TBE radar diagram (top right) summarising the stakeholder's last TBE assessment; as well as graphics for the Causal Chain and Impact on Supply (both to be explained in detail later). There are five text boxes that correspond to the BTRM categories. When signing into the system, the stakeholder's last entries in each category will be visible in these text boxes. When entering text into these boxes the whole screen is used and formatted according to the category fields shown in the Appendix. The basic layout for the dashboard is illustrated in Fig. 3.6 and the basics of data entry are described below. Note that the text in the text boxes here is for illustration purposes only and may not faithfully replicate the definitions in the ontology within the Appendix.

Capturing the 'Management Entries' of Change and New Business Needs

By management entry we refer to the Change and, optionally, any Business Need entries. The requirement is to capture any new information that constitutes a new opportunity or threat to the business. Change is defined as any event, business

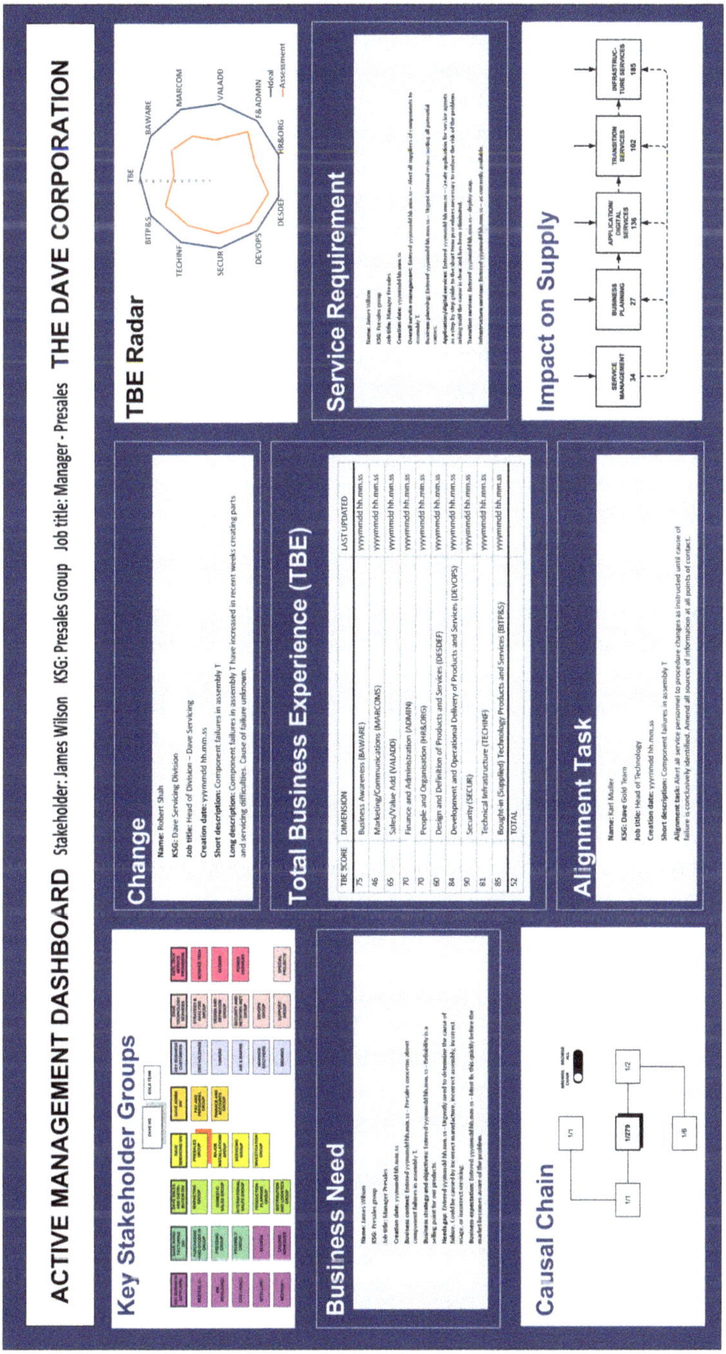

Fig. 3.6 The Active Management and Governance dashboard

activity or condition that has the potential to create a need for the business or service provider to do something differently, i.e. in order to maintain an equilibrium, retain or obtain a competitive advantage. Examples may include something imposed by shareholders, changed trading circumstances, competitor activity, a downturn in the economy, a pandemic which restricts the movement of people, or something new and relevant in terms of a technology, or a combination of these.

The new Change is entered into the Change box. The first time the Change is recorded a short name or short phrase is used to identify the Change that the stakeholder believes will (or could) impact the business. More detail is provided using the long description to describe the stakeholder's view of the Change. Multiple description entries against the one Change created by more than one stakeholder may exist over time as more information becomes available. In this way uncertainty arising from any ambiguity can diminish over time (offering a new approach to how we address the cone of uncertainty) as more becomes known about the Change and its likely impact. This and other Changes must be managed throughout the life of business and Changes that get missed in this way can be picked up at later stages when new Business Needs, or new TBE assessments are entered that reveal previously implicit Change.

Gradual Change may not be noticed at first, the classic being the effect of 'under the radar' competitor digital activity on market share. The condition may have developed over an extended period of time making the technology services provided gradually less relevant over this period. Other Change can be very sudden and difficult to anticipate such as a new entrant to the market or discontinuous product innovation, a fire in a manufacturing plant, a problem with a key supplier or key customer, a meltdown of the financial sector, or environmental disasters, etc. Some Change can be anticipated and monitored such as Change in market size and share as a result of marketing campaigns, forthcoming legislation, currency shifts, or incremental product innovation. Cumulative Change, or multiple causes having a combined impact on the business, can be masked from view and their cumulative effect can be unexpected. In these circumstances the more information that is available from as many perspectives as possible is very helpful in terms of disaggregating the causes.

Knowing when Change is taking place and reporting that promptly and appropriately to other stakeholders is very important. The research showed that organisations today report Change infrequently, probably because it's not a requirement of the IT methods. Where organisations have a culture of waiting until they can define a requirement before commissioning services, they are less likely to be able to maintain an adequate and timely service to their business stakeholders. Reporting change early so that discussion about the implications can take place may mean organisations changing the way they work and think about their business, and the technology that supports it.

Of course, knowing that a Change has occurred is unhelpful in terms of formulating a response unless the business implications of that Change are understood but it shouldn't be assumed that the stakeholder entering the Change will always understand what all of the implications may be. The senior stakeholder may be seeking opinion. Other stakeholders seeing the Change entry are able to enter notes indicating how this will affect them, in their roles, within their KSGs. These stakeholders will be prompted to supply any helpful information on the detail of the consequential Business Need, i.e. the Needs Gap created by the Change, any information relevant to the Business Context, changed Business Strategy and Objectives, and the Business Expectation.

The stakeholder's view of a modified Business Context can be expressed in whatever terms make sense to the stakeholder, e.g. it may consider how the need arose, the impact of a technology on internal and external value networks, or position the need with respect to past, current, and future states of the business, etc.

The stakeholder's view of Business Strategy and Objectives explains how the overriding aims of the business must be modified/have been modified as a result of the Change. This may be illustrated by focusing on the compelling components of the business plan perhaps referencing plant facilities and real estate, people, process, suppliers, products, markets, channel strategy, or technology.

The stakeholder's view of a Needs Gap explains what is additionally thought to be needed to achieve the Business Strategy and Objectives or to carry out the role. This may be a new/revised business process or capability, better data, a new service, or a technology or technical product/facility, etc.

The stakeholder's view of Business Expectation explains what kind of response is expected from those addressing the Business Need. Is the Needs Gap expected to be fully addressed reasonably quickly? If not should a short-term tactical/partial solution be considered and, if so, what aspect is the most important to the stakeholder or the business? Solution options might be suggested even though these might be rejected in favour of others later.

Not all of the information has to be provided, only what is known, thought to be likely, or is considered to be important enough to draw to the attention of colleagues.

In organisations where all stakeholders are able to contribute their ideas on Change and Business Need and for these to be visible to all, it would be good practice to first get the approval (outside of the control of the system) of the KSG manager or management at a higher level to ensure that the stakeholder's views were useful insights and did not conflict with business strategy. Having said this, the best ideas don't always come from the top of an organisation and every stakeholder will have a different perspective on the Change. Front line staff will often have insights that people higher up in organisations do not have and so democratising this activity can be useful and is one way of avoiding the HIPPO effect, i.e. where the Highest Paid Person's Opinion becomes the only opinion. If greater democratisation is seen as desirable, then such enterprises may wish to allow contributions to be entered without any requirement for higher approval and this would be a policy option designed to encourage collaboration. Where organisations are nervous about democratisation, then it simply has to set policies that correctly reflect the way in which it wishes to operate.

Capturing Assessments of the Total Business Experience

The TBE was identified as the Core Category, i.e. it accounts for the most variation around the management and governance of enterprise technology and reflects stakeholder satisfaction levels across all aspects of product and service from the TSP. It is not surprising therefore that the TBE should be at the centre of Active Management and Governance or its aim to provide a managed business experience. By continuously assessing the TBE we have a continuously updated view of value-in-use across the enterprise. The act of making assessments gives rise to value creation opportunities and together they bring service-dominant logic to the management of business technology. The TBE provides content and context in terms of when, who, where, what, and why. A TBE assessment as a component of a causal chain tells us if it was as a result of some explicit Change or new Business Need or whether it is an early indicator of Change that hasn't yet been articulated, or whether the assessment was created as a consequence of a recent failure in service, etc. In governance terms it also provides measures for alignment and agility. To achieve all of this the TBE must be capable of describing the many aspects of service and there must be a way of capturing the assessments.

Each assessment is a stakeholder's perception of the suitability of the TSP's products and services at that point in time. When a stakeholder is signed into the system the stakeholder's previous assessment of the Total Business Experience (TBE) is displayed in the central text box with the corresponding radar diagram in the top right graphic, and the Causal Chain of which it is a part in the lower left graphic. The TBE text box will show the stakeholder's scores for all 10 dimensions as identified by the full name or the mnemonic, the date and time that dimension was last updated, and a TBE total score as illustrated in Table 3.1. There will be options to view individual dimensions and the property scores and explanations given, to view the mean for the stakeholder's KSG and the group's consolidated comments.

The stakeholder is then able to enter new assessments for just those dimensions that the stakeholder believes need to be scored differently since the last entry. For data entry the assessment will switch to a full screen display where the full entry will be viewed and each dimension and the scores given to each property can be viewed in a similar fashion and optionally amended.

Attributes are not primitive codes but require qualification to provide meaning, i.e. a score to indicate the extent of the impact and a description by way of an explanation perhaps with examples of the difficulties this creates and what should be done. Thus, an explanation accompanies every assessment allowing an infinitely variable and flexible way of describing a service issue. By combining such explanations from multiple views requirements can be established and meaningful proposals for alignments tasks can be drafted. Without the text the TBE would require more classifications and there would be too many codes and still too few ways of explaining the problem. By providing explanations stakeholders are also justifying their assessments which makes it easier for them to amend an assessment when an alignment task has been successfully implemented.

Table 3.1 TBE summary as it appears in the TBE text box of the dashboard

TBE score	Dimension	Last updated
75	Business Awareness (BAWARE)	yyyymmddhh.mm.ss
46	Marketing/Communications (MARCOMS)	yyyymmddhh.mm.ss
65	Sales/Value Add (VALADD)	yyyymmddhh.mm.ss
70	Finance and Administration (ADMIN)	yyyymmddhh.mm.ss
70	People and Organisation (HR&ORG)	yyyymmddhh.mm.ss
60	Design and Definition of Products and Services (DESDEF)	yyyymmddhh.mm.ss
84	Development and Operational Delivery of Products and Services (DEVOPS)	yyyymmddhh.mm.ss
90	Security (SECUR)	yyyymmddhh.mm.ss
81	Technical Infrastructure (TECHINF)	yyyymmddhh.mm.ss
85	Bought-in (Supplied) Technology Products and Services (BITP&S)	yyyymmddhh.mm.ss
52	TOTAL	

Where entries from others within the group are seen, a stakeholder is able to 'agree' with the score and comments of others as appropriate and these will become part of the stakeholder's own assessment. It is possible to 'agree' with a comment but to submit a different score. Where a dimension and property score (out of 100) has changed, the reasons for the new entry are requested and totals recalculated. Assessments that haven't changed from the stakeholder's last entry need not be re-entered but are included automatically in the recalculation. As scores are entered the stakeholder can observe the effect of these in the TBE radar graphic. When the stakeholder is satisfied that the scores reflect the experience, the stakeholder will 'commit' which stores the assessment as a permanent record with a date and time stamp. Alternatively, 'cancel' will return the display to the stakeholder's unaltered previous assessment.

Where assessments are being downgraded the system will prompt for any known relevant Change to the market, technology, or service quality that may have given rise to new Business Needs and to this assessment. This may involve linking to an existing Change or to a Business Need entry or creating new ones.

The stakeholder will be asked to create the Causal Chain by creating the logical links between the notes about Change, Business Need, the TBE assessment and any Service Requirement to supplement the entry to ensure that cause and effect (or Change and response to Change) can be quickly understood (see 'Identifying service requirements', below).

The new assessment entry is retained and the mean for the KSG to which that entry belongs is updated with this latest entry, and all higher-level summaries are updated calculating a mean using the new data, i.e. at division and company levels.

External business supplier or business customer stakeholder entries will be accessible to the relevant business unit and optionally included in the business unit overall assessment (see 'Setting client policy').

The dimensions in Table 3.1 and their properties are based on the research, on observation, and on general practice, and are shown in full in the appendix but are described below. The stakeholder is asked to amend the score of each property as necessary out of 100 indicating how close it is to the ideal, provide an explanation for each new score given, in order to arrive at a new score for the dimension which will appear in the TBE summary and radar diagram. Amended scores will be dated with the date and time of the amendment in the 'Last Updated' column.

Every dimension will undergo change through the use of Active Management and Governance as agile business methods and 'Very Agile' become more widely used and stakeholders may choose to comment upon the speed of this transition as they submit assessments.

The dimensions of the TBE are described below:

Business Awareness (BAWARE) Describes, using seven properties (Knowledge of the Macro Environment; Knowledge of the Meso Environment; Knowledge of the Micro Environment; Knowledge of Key Performance Indicators; Knowledge of the Business Assumptions and its Strengths, Weaknesses, Opportunities, and Threats; the Budget for Business Awareness; the Leadership of Business Awareness), how the stakeholder perceives the TSP in terms of its knowledge of the business environment generally, the specific business sectors in which it operates, and the business itself.

Shortfalls in Business Awareness due to the separation of technology from the business will inhibit the service provider's ability to understand the Business Need and to deliver appropriate products and services. Strong Business Awareness enables the TSP to quickly respond to Business Needs, e.g. to provide answers without the need for research, to shortlist potential product suppliers thus saving time and money, or by generating ideas and strategies that can help move a business from being ordinary to being a leader. An understanding of the wider business context enables the TSP to have more fruitful discussions with the business unit managers which can enhance every aspect of service delivery. This applies not just to the TSP's knowledge of the primary business sector but where the business is diversifying where a knowledge of different practices, systems or technologies is needed.

Marketing/Communications (MARCOM) Describes, using nine properties (Brand; Products, Services and Their Markets; Communications Strategy and Stakeholder Communication; Internal Communications; Technology Strategy; Strategic Data Layer; Enterprise Architecture; Marketing/Communications Budget; Marketing/Communications Leadership), the stakeholder's perception of how effectively the TSP communicates externally with the business and internally with its own people.

The purpose is to build relationships, identify need, satisfy need, and retain the customer or consumer. In the case of technology services, however, this becomes more complex involving technical strategies and architectures. It is the means by which the services provider makes the business stakeholders, at many levels, aware of its understanding and the capabilities of its products and services so that buyers can get endorsement from their peers before a 'short listing' or purchase decision is made. The problem is that architectures are static and time consuming to build during which time events may have made them obsolete. The challenge is to find new ways of communicating this information.

If the services provider doesn't maintain an accurate picture of what the business is planning and wants, it can never be provided. If the services provider doesn't tell the business enough about its products and services and what they can do for the business, the business will have to come to its own conclusions and may go elsewhere. If the TSP's internal communication is flawed information is not shared and action doesn't necessarily follow.

Success at business awareness and marketing and communication will yield benefits to both parties at all levels but will most interestingly be assessed at the highest levels, i.e. at the board or executive level, because it is here where it is often at its weakest and only here that it is truly possible to take a cross-organisational or non-partisan view.

Sales/Value Add (VALADD) Describes, using six properties (Sales Strategy; Sales and Account Management; Pre-Sales Support, Innovation, Value Creation; Investment Portfolio, Order Book, Sales Pipeline; Sales/Value Add Budget; Sales/Value Add Leadership), how the stakeholder perceives the effectiveness of the TSP at matching its capabilities and solutions to Business Needs.

This will be assessed by the extent to which this is geared towards the business enterprise, the organisation of the sales team, the success of any pre-sales/sales activity, the co-creation of value by the business and the TSP through collaboration and innovation, and the nature of the output that may have been possible in terms of the investment portfolio. There may be a sales involvement throughout the enterprise's business planning, financial planning, prioritisation, and commitment stages. The TSP benefits by having a better order book and the business enterprise through better planning data, and ultimately from the value to be derived from the investments.

The sales process is about identifying the value that is being created by the offer to provide product or service and so persuade the buyer to enter into a financial transaction to purchase the benefit stream associated with the investment. This sales/value-add activity will operate at many levels but will ultimately be focused on the budget holder or divisional/business unit level where the benefits associated with an investment have to be considered.

Finance and Administration (F&ADMIN) Describes, using seven properties (Contractual Arrangements; Financial Management; Compliance; Admin Management; Growth Management; Finance and Administration Budget; Finance and Administration Leadership), the stakeholder's perception of the effectiveness of the Finance and Administration activities of the TSP, including all statutory, legal, and regulatory requirements.

This is key to the delivery of the core product or service in terms of ensuring that there is business confidence in the financial management and in defining what is being delivered, at what cost, under what terms, and in compliance of which laws. Stakeholders will reduce their scores when they experience mistakes, especially those which give rise to concerns about competence and viability. An early indication of these concerns allows corrective action to be taken before such views become widespread. The context, the property being assessed, and the explanation where the score has been reduced will usually provide the detail about what has to be done.

Financial and administrative management issues can cause as much pain and inconvenience for both the service provider and the business as issues with any other aspect of service. It is also an area where businesses will not expect to have a problem until, that is, one arises. Just as technical problems can become visible immediately across the whole the business, so too can commercial, financial, legal, and even administrative issues.

People and Organisation (HR&ORG) Describes, using seven properties, (Organisational Structure and Accountabilities; Culture and Motivation; Training and Development; Individual and Group Effectiveness; Recruitment; People and Organisation Budget; People and Organisation Leadership), how the stakeholder perceives the effectiveness of the TSP people, how they are organised, the culture of the organisation, and the HR management aspects of the TSP and the impact of this on the business.

Whatever we do and however we do it, we influence what people think about us simply by virtue of who we are and how we conduct ourselves. It has been said that people buy from people, that we associate with like-minded people, and that we only do business with people we like. Not only do we need to be able to trust people we deal with but in professional circles in particular there is plenty of evidence to suggest that we probably need to respect the experience, ability, professionalism, and the personality of those that we will choose to help us. This will influence how we organise ourselves in terms of responsibilities, lines of communication, how much trust we place in people, and how we relate to others. Although the focus here is the TSP, this is really a collective responsibility by both the business and the TSP because both parties stand to benefit from the relationship and it must be an issue for both at the highest level.

Design and Definition of Products and Services (DESDEF) Describes, using five properties (Understanding the Requirement; Product and Service Design; Build versus Buy Decisions; Design and Definition Budget; Design and Definition Leadership), the stakeholder's perception of the effectiveness of the TSP in establishing the business requirements specification, the design and technical specifications of the solutions, and the impact on the business. The Design and Definition of Products and Services will be assessed by stakeholders with different interests, including stakeholders responsible for the investments, the process owners, those who are using the solutions on a day-to-day basis, and those who could benefit if more functionality were to be added. All opinions will need to be taken into account.

The challenge here is that design and definition faults are more expensive to correct later in the lifecycle so the processes involved in these activities are important. These activities can also be time consuming and so by the time concerns are being raised about the output time is often running out. If they are seen not to be working by any stakeholder, this should not be ignored.

Development and Operational Delivery of Products and Services (DEVOPS) Describes, using 10 properties (Program and Project Management; Data Quality; Software Engineering; Deployment; Service Operations Management; Upgrades and Change Control; Support Services; Quality Management; Development and Operational Delivery of Products and Services Budget; Development and Operational Delivery of Products and Services Leadership), the stakeholder's perception of the TSP's acquisition or development of solutions and services, their deployment, operation and support services, and the impact of this on the business. This may be about delivering a new application, the business process re-engineering to support the change, or providing a support desk service. By combining software development (Dev) and operations (Ops) practices the aim is to shorten the systems development life cycle through continuous delivery with high quality software. The properties continue to make the differentiation between these different aspects of delivery so that there is every opportunity for stakeholders to be specific about concerns and the causes of failure.

Security (SECUR) Describes, using five properties (Site and Business Security; Information/Access Security; Business Continuity/Disaster Recovery; Security Budget; Security Leadership), the stakeholder's perception of the TSP's involvement in security provision when compared with industry best practice, legislation, or its effectiveness in protecting the assets of the business. Cybercrime and crime against service installations are increasing and ever more critical to the well-being of the business, its employees, and other stakeholders. Stakeholders may be aware of any slackness in controls long before they are exploited. An early warning of such a circumstance is useful, as are patterns of behaviour that may give rise to security exposures, or thoughts about which parts of the business may be vulnerable. Every report should be investigated with urgency.

Technical Infrastructure (TECHINF) Describes, using seven properties (Scope of the Vision; Skills Training for the Business; Introducing New Technology; Asset Register/Technology Deployed; TI_EEF, Ecological and Environmental Footprint; Technology Infrastructure Budget; Technology Infrastructure Leadership), the stakeholder's perception of the effectiveness of the TSP in maintaining the utility value of the technical infrastructure (hardware, software, and telecommunications) of the business and the impact of its activities on the environment. Scores may, for example, be reduced by stakeholders when new technology with a greater utility value or with less impact on the environment is not made available.

Bought-In (Supplied) Technology Products and Services (BITP&S) Describes, using 12 properties (Third-party Supplier's Business Awareness; Third-party Supplier's Marketing/Communications; Third-party Supplier's Sales/Value Add; Third-party Supplier's Finance and Administration; Third-party Supplier's People and Organisation; Third-party Supplier's Design and Definition of Products and Services; Third-party Supplier's Development and Operational Delivery of Products and Services; Third-party Supplier's Security; Third-party Supplier's Technical Infrastructure; Third-party Supplier's Supplied (Bought-In) Technology Products and Services; Third-party Supplier's Bought-In (Supplied) Technology Products and Services Budget; Third-party Supplier's Bought-In (Supplied) Technology Products and Services Leadership), the stakeholder's experience of using the products and services of third-party TSPs.

Note that the stakeholder's experience of the internal, main or only TSP, that is contracted directly to the business, i.e. the tier one, or sole, TSP has been captured using the primary dimensions of the TBE described above and will not use the Bought-In (Supplied) Technology Products and Services (BITP&S) dimension described here. This dimension is used to describe one of a number of other suppliers of technology products or services contracted directly to the business, or a product or service provider with whom a tier-one TSP, e.g. the main TSP, has a supply contract. This could be a third-party supplier of services, or a specialist software or hardware vendor.

Stakeholders cannot always be expected to know who supplies every service they receive but there will be lots of clues within an assessment. The supplier being assessed might be identified by the stakeholder's KSG (e.g. where a supplier is contracted to supply certain services to a specific group), the TSP's KSG (e.g. where a group has been created specifically to identify the TSP), the service provided (e.g. where a certain service has been outsourced to a specialist supplier or where a service has been outsourced to a specialist third-party supplier of services only for a given KSG), the textual description (i.e. identifying the product or service at issue) and by the description of what should be done, or by a trading name (i.e. by naming the supplier within the explanation). Other solutions are discussed in Chap. 5, The Automated Future.

Capturing the Service Requirements

Having entered an assessment of the TBE, the stakeholder may optionally suggest a Service Requirement that will address the problem that has been identified, or that the stakeholder believes is a logical next step which will work towards addressing the problem. As with other text boxes, when entering data, the Service Requirements text box switches to a full screen display.

The stakeholder is asked for a short description to describe the requirement and is then presented with five options: a choice of some action related to overall management of the service, some business analysis or planning exercise, some application/digital (development) services, some transition services to help to introduce new working practices involving technology, or infrastructure services to create the technical infrastructure to support new services. Whichever of these is selected, the stakeholder will also be asked to provide a long description by way of an explanation of what more precisely is thought to be required, or where attention needs to be focused. The Gold Team will later confirm this, add a man day estimate, and schedule any necessary work. Note there is no requirement for a requirements specification, just a little more information to help those who need to allocate resources.

The Gold Team may undertake some preliminary survey (the 'go and see' or *genchigenbutsu*) to establish what needs to be done to arrive at its estimates. Service Requirements will have been requested by other stakeholders and where there are similarities, e.g. where there are a number of stakeholder entries requesting Service Requirements raised in response to the same or similar issue but as seen from different perspectives, these may be merged and investigated at the same time by the Gold Team when considering alignment initiatives.

Establishing and Recording Alignment Initiatives

The Gold Team will have access to all of the entries as above. Some will be complete contextual narratives addressing the whole of the Causal Chain from Change, to Business Need, to TBE Assessment and will be requesting Service Requirements, i.e. the Change and a complete Spectrum of Business Requirement. Others will not be complete but in combination it will be possible to build narratives that can be validated during analysis, perhaps as some formal business analysis process or by analysing the entries cross organisationally using the browse facilities of the system. This will enable the Gold Team to plan an Alignment Task designed to address the issues being raised. This Alignment Task is linked to all relevant entries so that it is possible for stakeholders to identify what Alignment Tasks are addressing what issues and vice versa.

The Gold Team will exercise judgement about when it is appropriate to aggregate requirements if by doing so the problem solution becomes large, expensive, time consuming to deliver, and there is a commensurate risk of failure. Instead, they will consider disaggregation as described in Chap. 4, Analysing the Real-World Landscape'.

The Gold Team will deal with risk assessment, investment approval, and resources, etc. Although it is not envisaged that an Active Management and Governance system would be linked directly to the service catalogue, there is provision in this system to capture key dates, stage approvals, and the status of each initiative from the service catalogue so that all stakeholders will be informed of progress on all entries.

When Alignment Tasks are completed, there is an expectation that entries (including TBE Assessments) will improve and checks will be made to monitor this such that if this is not the case that an alert is raised. There may be an action to follow up with the stakeholders who raised the issue to see if the problem has been resolved (and to request a new assessment).

More About Value-in-Use

This chapter has provided an introduction into how Active Management and Governance can be used to start to change the storyline by using information earlier in the lifecycle whilst otherwise mimicking what we do today. The product and production paradigm for IT management has often limited what can be achieved through technology as was described in Chap. 1, The Need to Change the Way We Work. The present methods use a modern-day goods-dominant logic or production and manufacturing logic which focus on value delivery not value creation. By contrast, service-dominant logic, i.e. the logic of services operating in an end-to-end manner across communities, value constellations or value networks rather than value chains, can be conceived as systematic social innovation involving the continuous design and redesign of complex business systems. This is also facilitated by Active Management and Governance.

In applying service-dominant logic the customer, or stakeholder, is seen to be the value creator and the offering, i.e. product or service, is an input to a value-creating process. Thus, value is primarily defined in terms of value-in-use, from which, in virtually all cases in a market-based society, value-in-exchange can be derived and, as can be seen from this chapter, the key to value creation in the context of business technology services is the stakeholder.

Thus, Active Management and Governance introduces a service-dominant logic that establishes an overall encapsulating value creation concept to re-frame the production or manufacturing logic encouraging early collaboration, early benefits, and the democratisation of the creation of value to maximise the impact of business technology. The production or manufacturing logic is not replaced, just reframed, which helps to ensure we retain any value inherent in the product but add value through the true service element.

Active Management and Governance makes use of service-dominant logic both by continuously testing the value-in-use of the systems and services and through value creation, and the TBE concept is at the centre of this process. Thus, there is real value in stakeholders submitting regular assessments. By using these regular assessments, the business experience can be monitored and managed on a continuous basis involving stakeholders from across the whole enterprise, corrections made where necessary, and if the entry is an early warning it can be used for developmental or contingency planning. This systematic social and technical innovation enables the continuous design and redesign of complex business systems—something difficult to achieve by other means—especially those that were periodic, episodic, or spasmodic and involving just a few.

If service-dominant logic, and hence Active Management and Governance, is an attractive management philosophy internally, then it also has the potential to change the SLA by basing future such agreements on the TBE, Closeness of Alignment, and Agility of Response. This offers further insight into the meaning of value-in-use and value-in-exchange (Vargo et al., 2008).

In Chap. 4, Analysing the Real-World Landscape, we look into how Active Management and Governance can enable business technology to be managed, measured, and monitored, in ways that previously couldn't be contemplated. So far, we have looked at the functionality of Active Management and Governance as a serial process mimicking current methods with its static architectures. This next chapter introduces the concept of managing multiple parallel processes in a dynamic setting.

Reference

Vargo, S. L., Maglio, P. P., & Akaka, M. A. (2008). On value and value co-creation: A service systems and service logic perspective. *European Management Journal, 26*, 8.

Analysing the Real-World Landscape

4

Our thinking is often constrained by what has gone before and so this chapter explains how the business landscape can be analysed in ways that have not previously been possible. Over the last 50 years TSPs have focused on the discovery of requirements, the overall architecture, and the management of the development and operational delivery of products and services. This may have the appearance of being a thorough process but it is actually piecemeal or fragmentary and serial. The approach has its problems and the industry has tried to address this through the analysis of broad business areas to determine strategies and investment priorities. Sadly, these strategy studies were typically one-off initiatives which when left on the shelf quickly became out of date and were eventually unsuitable or even irrelevant.

The reason for this is that we aggregated requirements and used static techniques to map the dynamics of business. In a real-world context we have continuously to deal with multiple incidences of change across multiple functions each being affected differently depending upon their purpose. We need to understand how each part of the business is affected by change and using the latest information we have about the business respond using multiple parallel remedial processes. By constantly looking at what is happening external to the business, the emerging new business needs, and the experiences of the stakeholders across the whole of the extended enterprise, we can better understand what has to be done, e.g. we have an understanding of which parts of an organisation are most affected by Change, what is required, who is contributing and collaborating to business and system improvement, what demands are being placed upon the TSP, and whether, with hindsight, better decisions can be made in the future, etc. We are not just able to scale up what was done before, but we can analyse the landscape in ways that were not previously possible, and we are able to provide a management and governance process to monitor services to ensure they always meet the needs of the stakeholders in a timely manner.

In the past we created silos of thinking. We created business area silos, technical silos, customer or market silos, and even supplier silos. These in turn created stakeholder silos. With limited skilled resources we didn't often talk to stakeholders

D. Miller, *Managing Agile Business Technology*, Management for Professionals,
https://doi.org/10.1007/978-3-030-90598-9_4

across silos but within silos. That's what silos are for but it was easy to miss the consequences and opportunities associated with Change, particularly where they fall across the silo boundaries. Now, with a new management framework and the means to facilitate the collection of data potentially all stakeholders from across the business, TSPs, supply organisations and business customers, it is possible to consider how a broader, cross silo, view can be managed and exploited, whilst at the same time recognising the differences, and how this changes the way we work. Success depends upon new tools to provide insight that wasn't previously possible given our former limited vision and correspondingly limited capacity for analysis. In Fig. 3.3, we called this tool the 'Analyzer' and it would typically be available to the Gold Team but could be used more widely in conjunction with an Active Management and Governance Dashboard. That functionality is described in the sections described below.

Section one 'Real-world problems and solutions' sets out the complexities, ambiguities, and the dynamics inherent in what we try to manage and the data that needs to be collected, section two 'Defining the set to be analysed' describes how sets can be defined and introduces the concept of disaggregation to isolate causes and effects, section three 'Navigating the set' suggests ways in which sets of data may be viewed to provide insight into support across the enterprise, section four 'Hindsight' explains how the integrity of historical data can be maintained and how hindsight bias is eradicated, section five introduces 'Contagion' which helps us to understand how the effects of Change can spread through an organisation, section six 'Creating value' lists the various ways in which this management framework creates value including that from cognitive diversity, and section seven 'Active Governance' explains how Active Management and Governance provides an over-arching key control point for the provision of technology services across the business. The final section 'What is missing?' suggests that we are probably most concerned about what we don't know and this provides an introduction to Chap. 5, 'The Automated Future'.

Real-World Problems and Solutions

The real world is complex, ambiguous, and dynamic, and the same is true about gathering data from stakeholders about their needs and experiences, etc. It's complex because there's a lot of it and how it connects is often not known, it's ambiguous because there are other gaps in our knowledge and people describe things in different ways and from different perspectives and at different times, and it's dynamic because its continuously changing.

The Active Management and Governance dashboard is a sophisticated multiplex display that provides access to all of the information within the system in a manner which allows its complex relationships to be explored through the interoperability of its controls. The same dashboard is used by everyone though not everyone will have all of its capabilities. Its basics were described in Chap. 3.

The data entered via the dashboard from across the enterprise defines the Changes taking place in the business and technology environment and the consequential Business Needs, TBE assessments, and Service Requirements from all KSGs. The Gold Team has access to all of these entries on a continuous basis. Alignment Tasks are the Gold Team response to this changing picture of demand. The Alignment Task is linked to all entries that it is designed to address, either in full or in part, and stakeholders can see when these are entered and when they are due to be completed. When an Alignment Task is completed, affected stakeholders may also update their TBE assessments and link to the Alignment Task responsible for what is, hopefully, an improved assessment. The possible entries and their possible relationships can be represented as an entity model diagram, see Fig. 4.1.

All entries are dated and have a stakeholder name and KSG but none are mandatory, all can be added at different times by different stakeholders and links can be created at the time of entry or as the relationship becomes clearer.

This database of entries can be large and is an example of Big Data. Entries may employ block chain technology if tampering is an issue provided operational performance is not jeopardised. These different entries in the database combined with the dashboard features facilitate the capabilities outlined in the following sections.

Defining the Set to Be Analysed

The set of entries that is the focus of interest for analysis can be defined in a variety of ways, all via the management and governance dashboard. The primary definition of the set is defined by selecting the KSG of interest.

A stakeholder's own entries define the most basic set and is the default set for analysis. Stakeholders need to look back to see how they have previously scored the service so that their latest TBE assessment can sensibly reflect the latest events and any movement in service levels (improvement or deterioration). To aid this the graphic of the TBE Radar Diagram not only changes dynamically to reflect assessments as they are entered but also previous entries can be viewed by scrolling back and forth. The default browse priority is the TBE and we can browse assessment entries chronologically forwards and backwards, i.e. in the order in which they were entered. The graphic element to this is useful because the visualisation provides a mechanism to easily locate entries where the scores are of particular interest as we browse. Depending upon enterprise policy, browsing may always be limited to the stakeholder's own entries but there are other possibilities.

Five main ways in which the set can be defined:

- As a manager of a company, division, or group, the stakeholder can view all entries for that company, division, or KSG.
- Using the KSG, and with the appropriate permissions (see Chap. 3, 'Setting enterprise policy'), a stakeholder can view a whole group, e.g. a KSG, division, key customer, all customers, a key supplier, all suppliers, or a TSP, etc. This is achieved by selecting the appropriate box(es) on the KSG graphic.

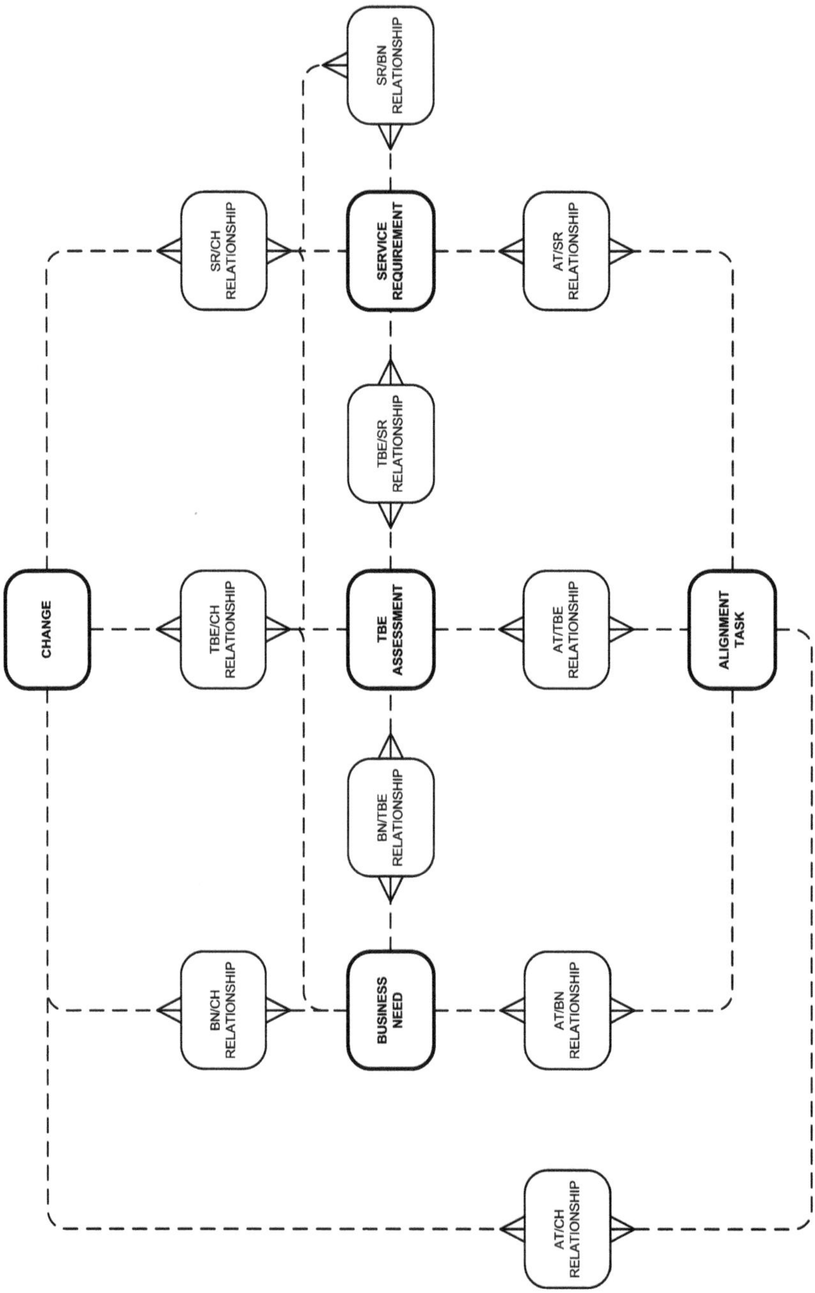

Fig. 4.1 Entity model of possible entries and possible relationships

- In conjunction with the KSG above, time parameters can be employed to set a date/time range to be examined, usually some point in the past and now, or around a past event of some importance.
- Using the Causal Chain graphic. A set can be defined by, or further limited by, selecting a category from the Causal Chain graphic and then browsing to select a specific occurrence of that browse, e.g. a specific Change, Alignment Task, etc. So, a set based on a specific Change would indicate the impact of that Change identified by the stakeholder group selected. A set based on a particular Alignment Task would indicate all entries for that group that had been taken into consideration wholly or in part in formulating that Alignment Task.
- A set can be defined by or further limited by using the Impact on Supply graphic. By selecting the service type that is of interest, e.g. application/digital development services, the set will be limited to entries that gave rise to requests for that service type.

Each of the above can be used in combination and incrementally at any time. So, for example, an analysis could be undertaken to look at a business division, this could then be limited in time, then by just looking at those entries that ultimately resulted in some kind of request for improved infrastructure services. The detail of how to navigate a set is explained in the next section, 'Navigating the set'.

It may be helpful to compare groups directly using the TBE Radar Diagram. This would simply require the standard radar diagram (Fig. 2.6) to have an additional plot for the additional group. If from the assessments it was necessary to drill down to look at the detail of a single dimension, e.g. Marketing/Communications, anyone with the appropriate authority could compare perceptions of the effectiveness of the TSP's marketing and communications of two or more groups and the extent to which assessments of the properties of Marketing/Communications fall short of what is expected or what would be ideal. Another way of using the TBE Radar Diagram is illustrated in Fig. 4.2 below. In this analysis the views of stakeholders concerning a single property, in this case Marketing/Communications, are compared across the whole of the extended enterprise including the internal and external business groups as well as the internal and external TSPs.

This analysis could be produced for any of the dimensions of the TBE selected on the TBE Radar Diagram. By comparing how all stakeholder groups perceive the TSP in terms of, in this case, its Marketing/Communications it's possible to identify potential issues. For example, here the TSPs rate their marketing and communications highly but the rest of the enterprise, especially manufacturing has a lower perception of the TSP's marketing and communications. It would be possible to explore this more fully by defining sets for further analysis based on such differences.

By examining sets and by comparing groups, it is possible to disaggregate what is reported. Through disaggregation it is possible to better understand the many different causes and effects and so work through solutions to what was previously hidden or complex. As in all things business-focused, we should recognise that businesses are affected by Change and that the many stakeholder groups will be

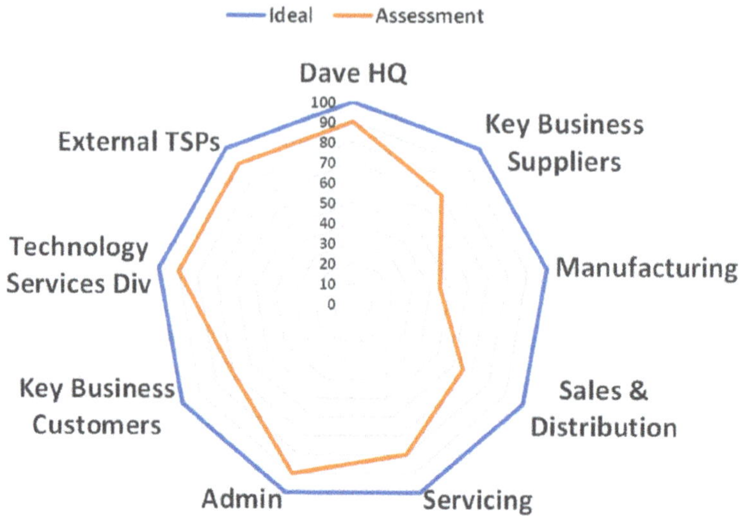

Fig. 4.2 Analysis of MARCOM across the enterprise

affected differently. By keeping stakeholder groups under review to pick up on emerging issues, the different needs of the different groups can be managed, i.e. in order to provide a managed business experience for all stakeholders.

Groups and individual stakeholders will have their own perspective on how their needs will be affected and navigating the sets allows us to disaggregate where appropriate. Every entry, including all assessments, has a business context. It has been created because something has arisen or the stakeholder considers that something could be improved. The context is the stakeholder's KSG and job title (and maybe role), the date and time of the entry, and what other entries it is linked to, e.g. a Change, new Business Need, or a TBE Assessment. If entries are from different groups undertaking different kinds of task, the contexts will be different and each entry could be describing the same problem from a different perspective, or different problems. It is important, by analysing the sets, to be clear about these differences, including those arising through cognitive diversity. The views of people with different perspectives, contexts, insights, experiences, and even different thinking styles enable organisations to avoid perspective blindness and group collective blindness to produce a more comprehensive picture of the requirement to take to the next stage (Sayed, 2019). It has been said that group processes by and large conspire to suppress the very diversity of viewpoints that we seek (Nemeth, 2018) and yet here we see that there is a management framework that gives time and space to creating value that can actually disaggregate and harness cognitive diversity rather than suppressing it.

By disaggregating cause and effect, solutions become more focused on smaller groups. Whereas previous methods tended to aggregate requirements leading to big solutions, this approach is likely to lead to many more small initiatives. Small initiatives provide more opportunity to focus on value creation, are easier to deliver, and more likely to succeed.

Navigating the Set

Having defined the set to be analysed, its content can be navigated so that entries can be examined in detail. This is achieved by selecting the browse priority to follow linkages of interest. To understand what can be achieved, it is necessary to understand how to set the browse priority, the mechanics of the browse, the role of the Causal Chain graphic in further limiting the scope of the set, and the function of the 'Browse Chain/Browse All' switch.

The basic mechanics of navigation are described in Figs. 4.3, 4.4, 4.5, 4.6, 4.7, and 4.8.

The priority of the browse (i.e. the primary Category being browsed) is set using the Causal Chain graphic in the bottom left-hand corner of the Active Management and Governance dashboard. The categories are represented by boxes which correspond to the positions of the text boxes on the dashboard. Note that the Alignment Task box is included though strictly speaking is not a part of the Causal Chain having been created by the Gold Team but it completes the narrative. As shown in Fig. 4.3, it defaults to browsing the assessments of the Total Business Experience (TBE) as indicated by that box in the Causal Chain diagram being highlighted. Assuming that the browse function is being used to browse the user's own personal TBEs this is showing the browse was stopped at the first of 279 TBE entries and this has a chain of one Change and one Business Need leading to the assessment and the first of two Service Requirements. It is also showing the first of six Alignment Tasks

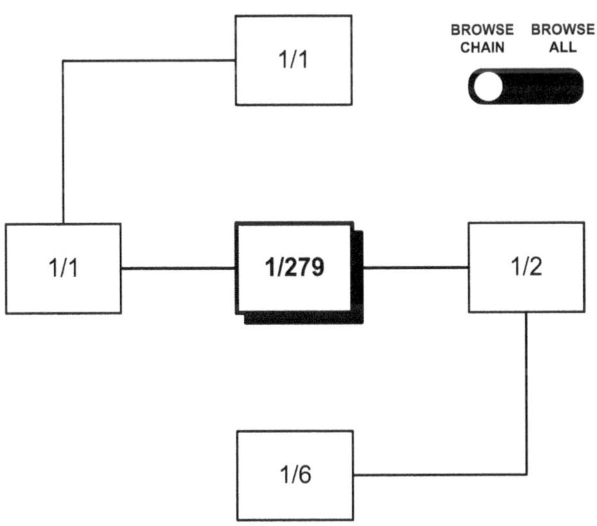

Fig. 4.3 Defaults to browsing assessments of the TBE

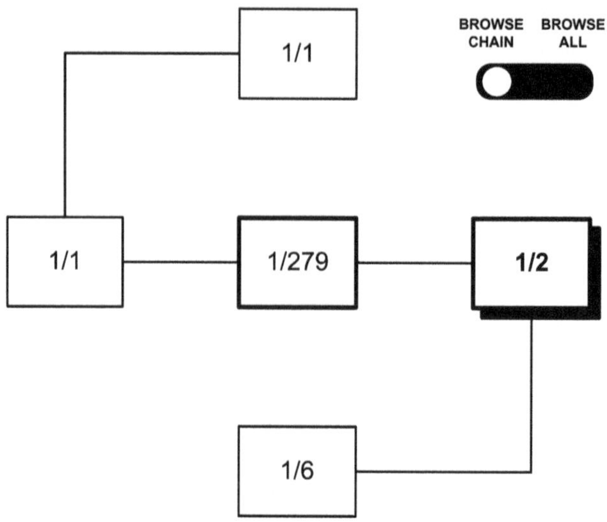

Fig. 4.4 Changing the browse priority to related Service Requirements

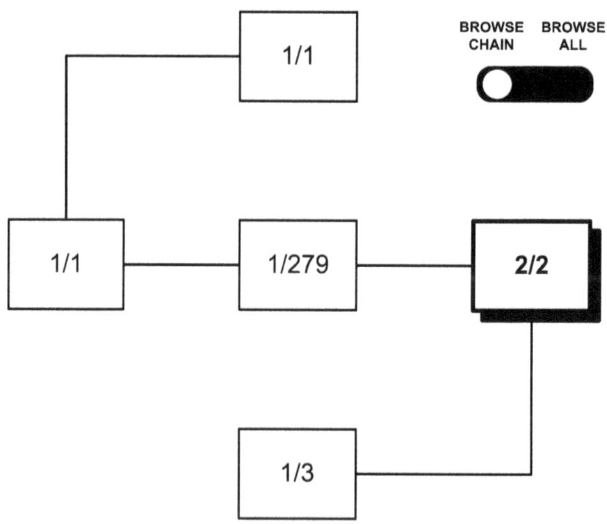

Fig. 4.5 As the browse is activated the sequence number Changes

resulting from the Causal Chain (these having been created by the Gold Team). Where there is more than one dependent entry as here in the case of the Service Requirements and the Alignment Tasks, the content of the first of these entries will initially be displayed in the dashboard text areas. The whole together provides the contextual narrative for the defined set.

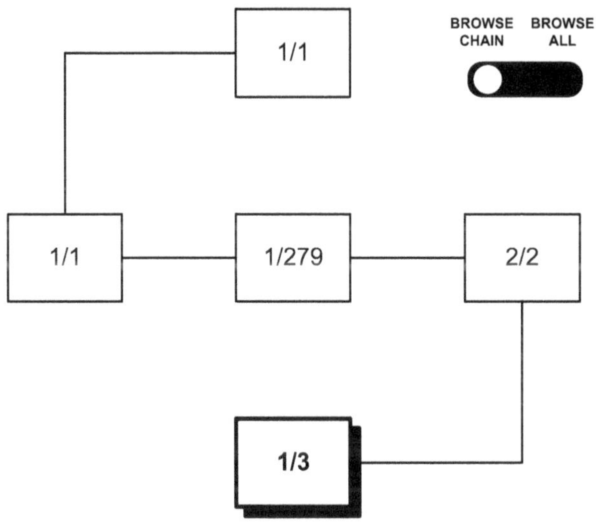

Fig. 4.6 Other connected entries can be browsed

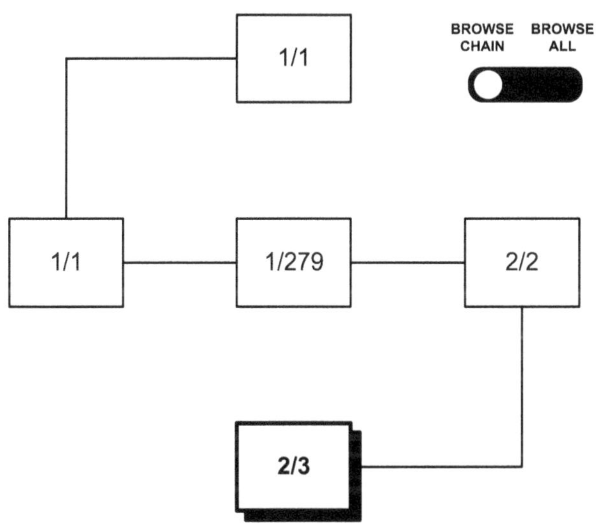

Fig. 4.7 The sequence number always shows which text entry is being viewed

A. To examine all related Service Requirements resulting from the assessment of the TBE being viewed, the Service Requirement box on the Causal Chain graphic is first selected as in Fig. 4.4. When the browse function is activated, the stakeholder is able to browse all Service Requirement entries in the chain (note that the switch on the Causal Chain graphic in this instance is set to 'Browse Chain').

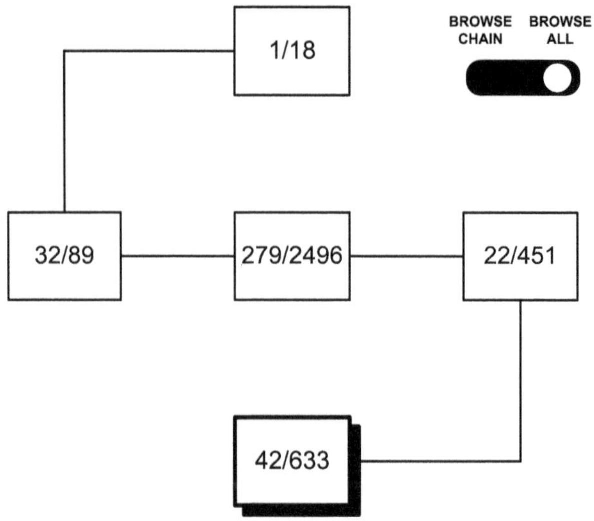

Fig. 4.8 The purpose of the 'BROWSE ALL' switch

B. As the stakeholder browses the occurrences of Service Requirements, the number Changes in the box on the Causal Chain graphic to indicate which entry, in sequence, for that chain is being viewed in the Service Requirement text window as in Fig. 4.5. Note that there are now three Alignment Tasks associated with that Service Requirement and the Alignment Task text box will now be showing the first of those.

C. By changing the browse priority now to Alignment Task, we can browse the three occurrences of that Category that are linked to what in this example is the second Service Requirement within the chain being examined as in Fig. 4.6. Selecting Alignment Task on the Causal Chain graphic activates the browse function to operate on the Alignment Tasks allowing the stakeholder to browse all such entries in the chain and view them one at a time in the Alignment Task text box.

D. As the stakeholder browses the occurrences of Alignment Task, the number Changes in that box on the Causal Chain graphic to indicate which entry, in sequence, for that chain is being viewed in the Alignment Task text window as in Fig. 4.7.

E. If at any time the stakeholder moves the switch on the Causal Chain graphic from 'Browse Chain' to 'Browse All', the graphic will now display the total number of occurrences of each category in the population available for analysis. The entry being displayed doesn't Change but its sequence number is its occurrence in the whole population as in Fig. 4.8. The stakeholder can now browse chronologically

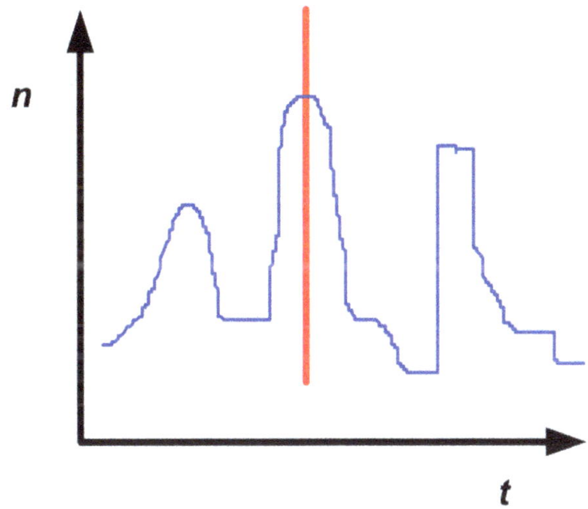

Fig. 4.9 Browsing a timeline of entries

backwards and forwards through the population choosing any of the categories and follow the Causal Chain linkages that have been created to provide the stakeholder with an insight into all related contextual narratives.

It is helpful if the Causal Chain graphic incorporates browse controls which include the FFWD/FBWD (fast forward/back), FWD/BWD (forward/back), STEPF/SEPB (one step forward/one step back, II (Pause), and □ (Stop) features. When used to browse the TBE entries, the TBE Radar Diagram scrolls chronologically through the chosen set and the occurrence on the Causal Chain graphic increases or decreases according to the direction of the browse. Thus, the TBE entries can be browsed rapidly, slowly, or one at a time in any direction until the point in time that is of interest has been reached. Note that text in text boxes is not displayed whilst the browse feature is in use, instead there will be a timeline showing a graphical representation of the number of entries in the set or population together with a line indicating where in the set or population the browse is positioned at any point in time, see Fig. 4.9. This line, shown in red in the figure, appears in the text box and moves in conjunction with the other dynamically changing graphics during the browse process. In this way the stakeholder is aware of the location of the browse relative to the population of entries and could easily find those times of most activity and perhaps what events had prompted them.

Other useful browsing features can be incorporated into the dashboard functionality. For example, there might be a message that warns the stakeholder when a browse moves outside of the context of the original Causal Chain (and asking if the stakeholder wishes to continue). Thus, having explored the current set the stakeholder could continue the analysis by moving on to look at entries made at a similar time but not necessarily connected to the current train of thought. When the browse is stopped the Causal Chain graphic displays the number of linked Categories for the TBE assessment currently being displayed in the TBE text box. The linkages then provide the contextual narrative for that assessment and the means of compiling an audit of the response. Each entry is unique as defined by its date and time stamp and stakeholder details (see Chap. 3, 'Knowing the stakeholder').

There are many other browsing possibilities that will facilitate further analysis. Some examples are offered in the paragraphs that follow to illustrate the power of these navigation features as a general management tool.

A group manager will see the group average and may also undertake the same kind of exploration as above for every entry in the group in date and time sequence. This same feature would allow divisional managers to compare and explore divisional assessments perhaps by specific KSG in order to determine where problems were arising and why. The Gold Team, on behalf of the CEO/Board/Executive, might compare and investigate assessments across the divisions perhaps before focusing on a particular division or KSG, to formulate responses in the form of Alignment Tasks that were optimal for the business. Clearly in these cases the logic of the browse is the same but the size of the set being searched is the variable. If the organisation was a global business, interests would be broader and the same browsing principles and features would allow managers or analysts to gain an insight into how well services meet the needs of the various parts of the business, including its supply chain and customer base, right around the world. There may be alerts, or problems may come to light during such explorations, and these might lead to further more detailed investigations, e.g. as in *genchigenbutsu* or the 'go and see' principle of the Toyota Production System.

To facilitate the comparison of the divisional assessments as in the example cited above, the TBE Radar Diagram could be made to be capable of displaying multiple plots but otherwise work in just the same way, i.e. by displaying successive assessments by scrolling through a defined population or set in date order. To search a large population for a set of conditions then this can be done by selecting the category as the browse priority for the set and view it using a time lapse feature when each entry will be displayed in succession in sequence at a speed chosen by the viewer. Alternatively, the set can be browsed backward and forward on any of the categories that are of interest by selecting and highlighting the appropriate box from those displayed on the Causal Chain graphic and then using the appropriate browse controls in the graphic. The Causal Chain graphic further assists this choice by constantly updating the number of each type of category linked to the browse entry during the browse operation. In this way the whole of an Active Management and Governance database can be navigated and any Causal Chain can be explored.

Maybe the browsing stakeholder is the CEO or other senior person in the company and will want to review the business Change entries and monitor the company's reactions to these to assess the impact and likely outcome. By selecting Change as the browse priority, that stakeholder could browse all Changes using the scroll buttons that appear on the Causal Chain graphic when this feature is selected. Having found the Change that is of interest that stakeholder could then switch the browse priority to Business Need and browse through all incidences of Business Need that have been entered that are linked to the selected Change. By doing this the stakeholder is able to explore the impact of a Change both in terms of the needs arising and, from the resulting Impact on Supply graphic, the resource implications. The stakeholder could go on to explore the detail of the TBE assessments and Service Requirements linked to the Change and Business Needs in order to understand the impact of the Change across the divisions and KSG's. Perhaps the CEO is interested in assessing these reactions in order to gauge the organisation's readiness for such a Change. Ultimately, this information would be used in discussion with the divisional heads to allocate budget, shape the eventual Alignment Task(s), and facilitate governance by allowing the CEO to monitor the closeness and agility of services across the extended enterprise (see the later section in this chapter, 'Active Governance') as the Change takes place.

A Gold Team member or a services manager for a TSP may want to scope out a project, assess future workload, or provide cost estimates. The Gold Team member or services manager would perhaps want to limit the set to be analysed by confining it to a division of the business by using the KSG graphic and then explore the Service Requirements associated with a particular Change. Or perhaps the need is to look at the total demand for Business Planning (e.g. by using Impact on Supply graphic and selecting the Business Planning services box) and examining the assessments, etc. that have given rise to these, or use this information to build a contextual narrative around a proposed Alignment Task for that service type based upon the TBE assessments, Business Needs, and the Changes to which these are linked.

The four graphics can be used in conjunction with one another to observe past or current activity. The normal focus when the TBE has the browse priority is the TBE Radar Diagram, however, whilst browsing backwards and forwards on the TBE the KSG graphic will also be changing to reflect the shape of the organisation at those points in time. The colour coding of the KSG provides a quick and very visible way to detect when changes take place to the organisation, i.e. KSGs added/removed or reconfigured, enabling the stakeholder to explore entries that were made around that time. Setting the browse priority to Business Need, Alignment Task, or Service Requirement and browsing whilst observing all of the graphics around the moment the KSG chart changes would also be an interesting perspective for analysis.

If Business Need is selected as the browse priority in the Causal Chains graphic, scrolling back and forth allows similar needs to be identified within the set selected. This may be useful as the same need can be expressed in a number of different ways and this provides a mechanism to explore the differences. Again, the number and nature of the linkages shows where any further more detailed exploration may be most fruitful.

If Alignment Task is selected as the browse priority in the Causal Chain graphic, scrolling back and forth on this category allows an examination of the needs that are being addressed by each task, within the set that has been selected. Doing this for completed tasks allows the assessments to be examined to determine whether expected improvements are being achieved. This may give rise to follow up actions, i.e. to determine that improvement has been made, as input perhaps to a post implementation review.

If Service Requirement is selected as the priority in the Causal Chain graphic, scrolling back and forth on this category allows an examination of the reasons that

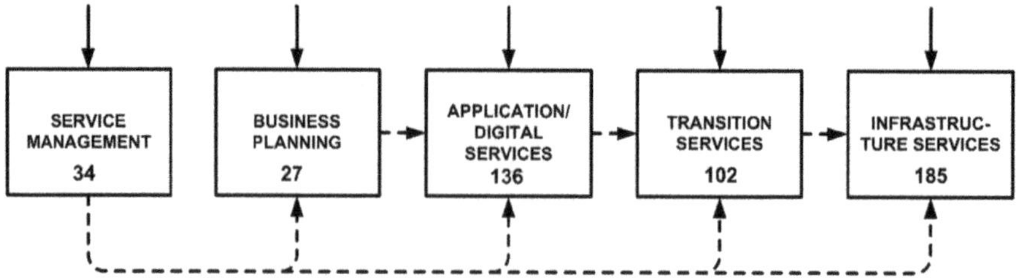

Fig. 4.10 Impact on Supply graphic

gave rise to the requests, within the KSG group selected, and the actions being taken. Whilst browsing backwards and forwards on the Service Requirement category and observing the Impact on Supply graphic the stakeholder can see how the demand changes over time and this helps with resource planning.

The Impact on Supply graphic, see Fig. 4.10, within the Active Management and Governance Dashboard, always provides a quick overview of how an individual or set, as defined by the KSG component and other settings, is impacting on the services, e.g. for the set defined by a given Change and a particular Division or Group. This is useful for viewing those entries that have given rise to the needs of a particular service type. In this way it is possible to determine which Changes or Business Needs are prompting that demand and this can be explored further using the browse features already described. This is also helpful when deciding whether to aggregate or disaggregate when creating an Alignment Task in a particular part of the business because some can be implemented immediately whilst others will require some investigation/planning. Note that the numbers in the Impact on Supply graphic always reflect the set and browse settings so when considering the total resource implications, the set must be defined appropriately.

The Active Management and Governance Dashboard is a sophisticated multiplex display that simplifies the complex relationships being managed. As the defined set, or whole database, is being navigated we have seen from above that the whole dashboard is kept in sync with the browse. Its role in terms of data capture and the interoperability of controls and displays is summarised in Fig. 4.11.

The shaded boxes focus on the graphic and the unshaded boxes refer to inputs and controls of various kinds. Understanding this summary of what has gone before will be helpful when reading the remainder of this chapter. In very large enterprises it may be desirable to add the timeline distribution feature to the browse function of the other categories, i.e. in addition to the TBE, to establish where in the total population of the set the browse is currently located.

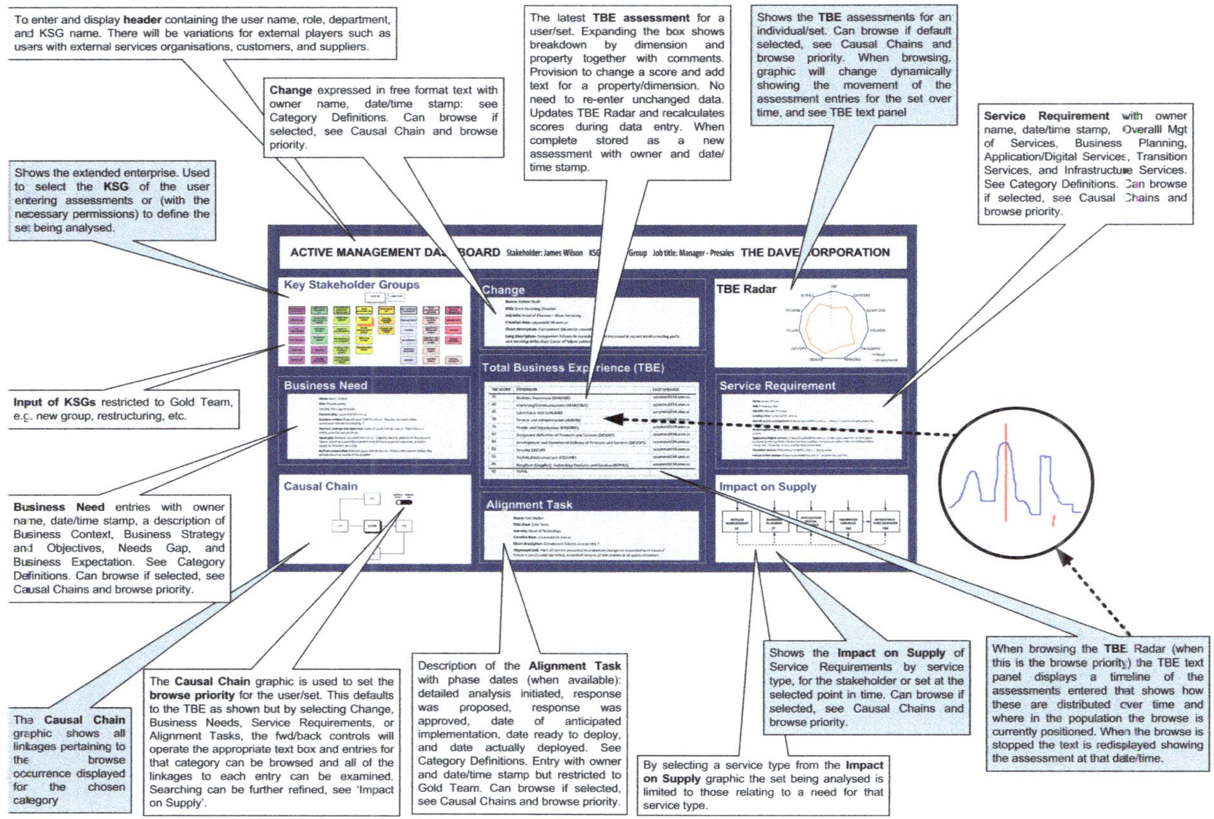

Fig. 4.11 The interoperability of controls and displays

Hindsight

Hindsight is said to be a wonderful thing but as we know there is a common tendency for people to perceive events that have already occurred as having been more predictable than they actually were before the events took place. This is known as hindsight bias. It was William Blake who said that 'Hindsight is a wonderful thing but foresight is better, especially when it comes to saving life, or some pain!' The hindsight features of Active Management and Governance similarly help us to better understand our past decisions, and so to improve our foresight, and to avoid pain.

As described earlier, assessments of the TBE are continuously amended up and down to reflect the circumstances that the stakeholder is experiencing and over time a contextual narrative is provided by constructing a Causal Chain by linking each assessment to any of the other categories of Change, Business Need, and/or Service Requirement that may apply. Thus, a Change can be notified and stand-alone but it will probably end up being the first category of a Causal Chain of entries by various stakeholders to explain a new and consequential Spectrum of Business Requirement and, at a later date, responses will arrive in the form of Alignment Tasks, and then hopefully more assessments showing improvement. In this way the contextual narrative develops.

Even though these contextual narratives help us to understand more about assessments and what should be done, history tells us that the way in which we behave is not bound by deterministic laws. History cannot always be explained deterministically and the future cannot be predicted because the world is chaotic (Harari, 2011). Indeed, in some cases predictions can even change the outcome. We could say that the present is the intended or unintended result of history but nevertheless we are still interested in what action was taken in the past and why. Hindsight is important if we wish to understand more about the realities of an historical situation as it played out, the reasons for past actions, the consequences of past actions and the possibilities going forward or in similar future situations. Events can be misinterpreted and if we are to derive maximum benefit from the history it is important that misleading entries in the present are not left but are corrected once better information becomes available, and, equally that past entries deemed to be invalid in this process are not hidden or deleted. Thus, it is important to maintain the full integrity of the historical data in order to understand the context for the original action as well as the context for subsequent corrective action.

A Change entered historically may have been an assumption about the future which, with hindsight, may now be seen to have been invalid. It may be that Business Needs identified as a result of the Change that were entered are now also seen to be invalid. Indeed, Service Requirement resources may have been allocated and even Alignment Tasks scheduled for deployment predicated upon these Business Needs. Whole Causal Chains may have been built on invalid assumptions and these will be misleading or even costly and so action must be taken in such eventualities.

In Active Management and Governance, the overriding principle is that nothing is deleted from the database; we can only change the status of the existing entries or

Fig. 4.12 Hindsight1—the original entry

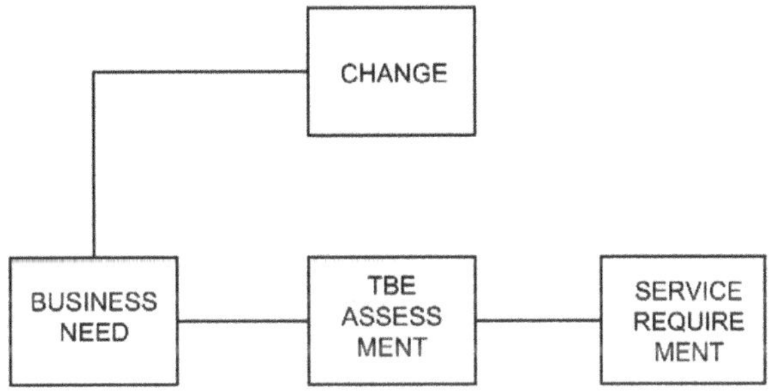

Fig. 4.13 Hindsight2—incorrect Business Need marked as invalid

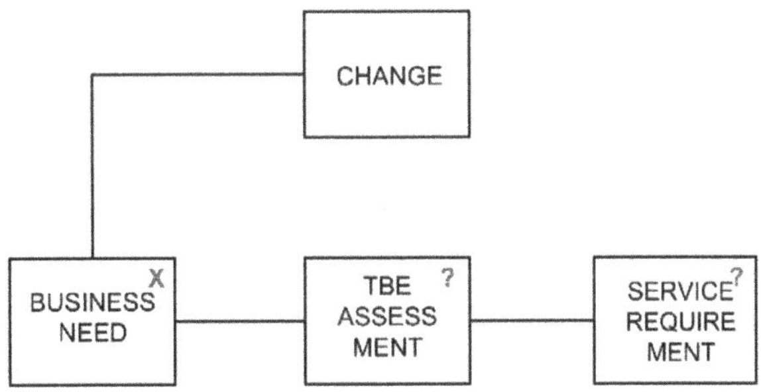

add new entries. This enables us to enter information in a timely manner and correct, or refine, this when more information becomes available. Thus, we can always look back at a past situation, e.g. to one where the Change or Business Need was assumed to be correct, in order to understand with hindsight our past actions, whilst simultaneously entering more recent and more accurate information for actively managing the situation going forwards.

To prevent stakeholders building upon entries known to be invalid and yet preserving the reasons for past actions, we must derive a way of retaining both the incorrect and the corrected information. To illustrate the approach, we will take the example shown in Fig. 4.12 which shows the original Causal Chain entry.

We indicate a change to the status of an existing entry by marking entries that were later found to be incorrect with an ‘X’. This could apply to a Change, Business Need, Assessment, or Service Requirement. In the example, it is the Business Need entry that with hindsight we consider as being invalid and so we mark it with an ‘X’. We would then mark all dependent categories in the remainder of the chain as ‘?’, see Fig. 4.13. Note that nothing is deleted. All items marked ‘?’ must be checked by the owner to assess whether they are still valid (by removing the ‘?’) or are indeed invalid (need to be marked ‘X’). Again, nothing is actually deleted and, like the original entries, all Changes of status to ‘X’ and ‘?’ or their removal are dated thus preserving the audit trail of all activity.

New information can now be added. In the example the Change remains the same and the stakeholder adds a new interpretation of the Business Need, in this case even

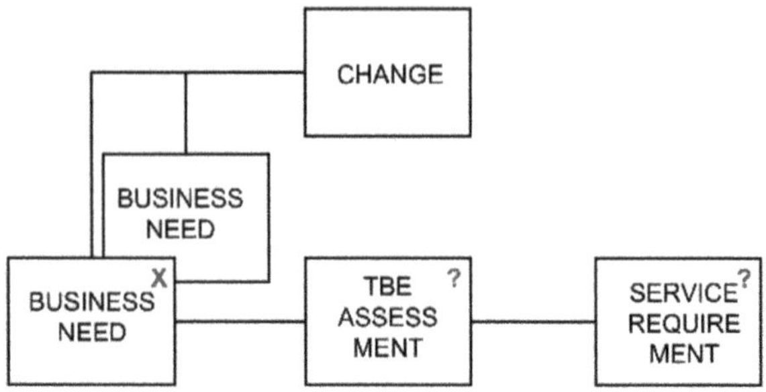

Fig. 4.14 Hindsight3—inserting a new Business Need

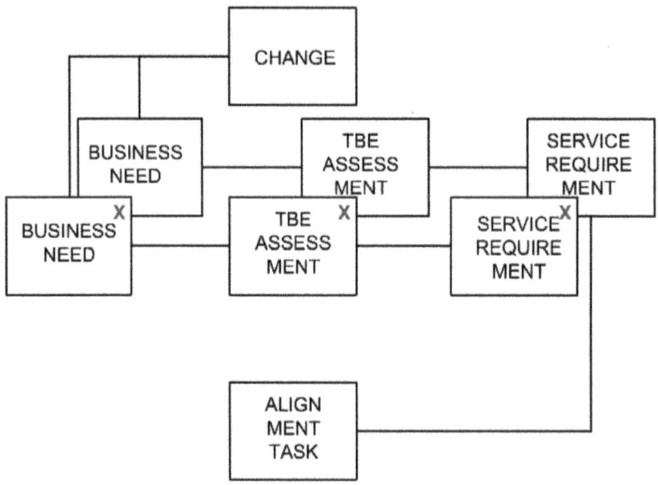

Fig. 4.15 Hindsight4—checked categories marked as invalid but retained as history

before the entries dependent on the incorrect Business Need have been checked, see Fig. 4.14.

By adding the new Business Need we start a new chain allowing new TBE assessments, etc., to be linked to this as appropriate. In the example, the queries (?) have now been checked and marked as invalid, see Fig. 4.15.

Note that the system marks all dependent categories in a chain as '?' automatically when a higher-level entry is marked as invalid. The owner of an assessment marked '?' then has choices as follows:

- If the owner believes that the assessment is still valid even though the higher-level entry in the chain to which it was linked is now invalid, the owner can remove the '?', optionally link to any new or existing valid higher level in the Causal Chain, i.e. another Business Need or Change, and re-commit remembering to check any

dependent Service Requirements by removing the '?' if still valid or replacing the '?' with an 'X' as below if it is now invalid.
- If the owner believes that the assessment is now invalid because the higher-level entry in the Causal Chain to which it was linked has been marked as invalid, the owner will mark the assessment (and any consequential Service Requirement) as invalid 'X' as it was based wholly on an invalid assumption. If there is any ambiguity after the changes, the stakeholder may choose to enter a new assessment.

Remember that all previous entries (valid and invalid) are retained and all actions have a date and time stamp within the contextual narrative thus retaining the complete history of what happened and when. When browsing categories, the database will show each type of category in date order including whether they are valid, invalid (X), or in need of checking(?).

Once a Causal Chain has been modified in this way the impact on supply will automatically adjust but routinely stakeholders and management must be sure that the following apply:

- That Causal Chains have been examined to ensure that all dependent entries marked '?' have been checked.
- That any Alignment Tasks that are active and actioned to deliver against requirements resulting from invalid assumptions are challenged and if necessary cancelled/modified.
- That current assessments have narratives that have integrity require no further explanation and eventually reflect the fact that threats or opportunities have been addressed by an alignment task, i.e. that TBE Assessments and Service Requirements are not outstanding as a result of the corrections and that there is an eventual Alignment Task where value ideas have resulted in positive outcomes.

By maintaining the integrity of the historical data in the manner described above the database always reflects past and present views of reality thus offering many more possibilities for value creation, management, and governance.

Contagion

If hindsight is helping us to maintain the integrity of past data, Contagion is using that past data to help us to anticipate the future—it can provide foresight. Conventionally contagion is used to describe how something spreads from one place, or person, to another. In the context of Active Management and Governance, Contagion shows how the impact of a Change spreads throughout an organisation from the early signs that a Change is happening to experiencing the full effect. Some groups will respond quickly with new Business Needs, perhaps because they see that they will be immediately affected, whilst others may not at first see any significance in the

Change. The time it takes for a group to work out the impact of a Change in terms of the Business Needs, the TBE, Service Needs, etc. affects its Contagion rate across the extended enterprise. Active Management and Governance will show us how the recognition of the needs arising from different kinds of Change has spread through the organisation.

This is helpful in anticipating or predicting the impact of similar Change in the future. If the future impact can be anticipated or predicted based on past events, our foresight is improved. This means that we can provide solutions sooner to the parts of the business likely to feel the effects first, and we buy time to ensure that those solutions can also be scheduled to meet the needs of the rest of the business in a timely manner whilst maintaining alignment and agility throughout. Every organisation will be different but it's clear that taking some time to understand the characteristics of Contagion, particularly in large and complex enterprises for given types of Change is likely to be beneficial. The task of understanding Contagion will initially be with the Gold Team, on behalf of the rest of the enterprise, where it can be tested out as opportunities arise.

Creating Value

What has previously stood in the way of any value creation process were the IT methods for delivering solutions and their dependence on burdensome IT procedures that inhibit mindfulness, creativity, and innovation—and hence value creation.

In contrast, Active Management is seeking to understand where business technology products and services do not meet the needs of the business and how this can be addressed. Ideas for improvement are sought via every stakeholder entry, i.e. the Change, a new Business Need, a TBE Assessment, or via a Service Requirement. These will not necessarily be detailed but will be targeted, i.e. will directly focus on the aspect that is deficient and explain the reason for the stakeholder score given. The TBE comprehensively challenges the service and the stakeholders are prompted to comment on all relevant service gaps. The stakeholder, having made an assessment and left various explanations may also have recorded a Service Requirement which the stakeholder will have considered to be the next step in addressing the Needs Gaps expressed in the TBE entries in order to fix the problem. Each entry was recording some aspect of value, i.e. how value has been reduced and what can be done to restore it or create new value. The process encourages these to be reported early, e.g. where Change may have occurred, and importantly, by making this information visible to others across the enterprise enables everyone to contribute, innovate, and collaborate which has a multiplying effect on the resulting creativity—and value.

It has been explained that regular monitoring of the TBE is a good mechanism for continuously testing the value-in-use of the systems and services. It describes the many aspects of service and provides measures for alignment including the intellectual as well as the abstract, or social and cultural, dimensions. To maintain a managed business experience, it is necessary to assess all of the data continuously and create Alignment Tasks as appropriate in order to deliver this value that has been created.

Past methods have neither given the time nor space to value creation but in Active Management, as in the real-world landscape, it is central to the ethos. What we have discovered in this chapter is that there are new ways of creating additional value by using the features of Active Management by enabling us to learn from what has already happened—even in very complex organisations. Managers can easily define sets of interest for analysis, sets can be navigated around events of interest, by the disaggregation of entries small low cost and low risk solutions can be implemented at speed, by understanding what Hindsight can tell us about past actions we can improve future decision making, and by understanding Contagion we can be better prepared for the future.

This section has explained how, by changing the way we work, we can bring the management of business technology up to a level with other experience-centric disciplines and industries with a defined business experience. It has shown how we can introduce a service-dominant logic that establishes an overall encapsulating value creation concept to re-frame the production or manufacturing logic. It has shown how to encourage early collaboration, gain early benefits, and instil the principles of the democratisation of the creation of value to maximise the impact of business technology.

All of this now places a responsibility on the governance processes to demonstrate the continuous effectiveness of the services.

Active Governance

If the Governance of business technology is failing most businesses and if it is to involve the business more, address a broader spectrum of business needs, connect with business more effectively, and at more levels, and for this to be a continuous process (see Chap. 1, section headed 'What we should know about the management of business technology') then the question is, how?

As has been said by many, at the end of the day, it is what the customer, or stakeholder, thinks and says about services, and service providers, that matters. However, TSPs have never had a complete picture of what stakeholders (even important ones) are thinking and saying about them, or how well they are doing, or even a complete picture of what they were supposed to be doing. Active Management and Governance puts the business back in control through Active Governance. It is always active and is managed by the board of directors perhaps with the assistance of the Gold Team. In Chap. 2 section, 'The BTRM as a governance framework', it was explained that the two themes of the BTRM are 'Closeness of Alignment' and 'Agility of Response'.

The first of these, 'Closeness of Alignment' in governance terms, is discussed below:

- Using Active Management and Governance it is possible to monitor the Closeness of Alignment of technology service provision with the needs of all parts of the business both before and after a response to Change and on a continuous basis. If there are no outstanding Business Needs, if all TBE Assessments (across all properties) are close to the ideal, and if there are no outstanding Service Requirements, then the technology services can be considered to be aligned with the business, i.e. closeness is a measure of alignment. Furthermore, if this is not the case, the impact or importance (from the score) and what is needed will have been explained and this can be examined in some detail in the manner described, so that follow-up action can be initiated.
- The degree of Alignment is expressed by potentially every stakeholder submitting assessments and these are being submitted on a continuous basis. There may be many reasons why a stakeholder is moved to submit an assessment, e.g. to express a view about poor performance, to register the effect of a new requirement on the job role as a result of a new Business Need, to note an important improvement in service after a new facility was implemented, etc. Because the assessment is always scored relative to what the stakeholder considers to be ideal, we also have a measure of the Closeness of Alignment and this can be viewed for each group, each division, and if needed for key suppliers and key customers. By inviting service providers (internal and external) to participate in the process, we can also compare these views with those of the business stakeholders and gather insights into why any differences in perception throughout the business enterprise might exist, what steps are being taken, and just how successful we have been so far in addressing these.
- Assessments address around 80 properties classified into 10 different dimensions. These include of the important areas of Governance including ensuring that all aspects of service have the necessary leadership, budget, and resources to address the business priorities. Even a good average score can mask serious deficiencies within the detail and hence the importance of the degree of disaggregation that this process introduces.

The second theme of the BTRM, 'Agility of Response' is described below and it is important to say that it should not be confused with 'Agile' as used to describe in very general terms a fast-track development methodology for bypassing detailed requirements specifications:

- Here it refers to the Agility of Response to Change to ensure that the management of business technology is as agile as the technology itself. Agile development was only concerned with the time from receiving a requirement to the time the solution is delivered, however it did also recognise the value of early and continuous delivery of software and the need to break down large projects into smaller deliverable parts. It must be recognised though that many TSPs are still reluctant to act until they have a complete requirements specification and IT management had no reliable way of ensuring the early notification of Change.

- Active Management and Governance is able to detect the first signs of an emerging new requirement from observing entries for Change, New Business Needs and TBE assessments and the Agility of Response is measured from those first signs to the time the solution, i.e. any related Alignment Task, is delivered. It is a true measure of the Agility of Response and therefore it is in the interest of the TSP to help the business to work out what is required as soon as possible.
- Active Management and Governance has coined the term 'Very Agile' to explain an approach to Just in Time delivery (JIT) of business technology solutions as SOA governance and features of Active Management and Governance come together to allow development work to be undertaken before the full requirement is known. This can be further enhanced by using the features of Contagion for predicting future impact and Hindsight to learn from past actions. 'Very Agile' is explained further in Chap. 6.

There are some rather more general points that have to be made about this key control point of Alignment and Governance:

- By closely monitoring the successes associated with the Alignment and Governance processes across the full scope of the relationship between business and technology (as defined by the BTRM) the risks can be re-assessed in order to learn from those experiences. In this way, perhaps as never before, it can be established how the processes can be improved. Risk assessment is also a feature of 'Very Agile' where it takes on a different significance as is explained in Chap. 5.
- As Active Management is technology-agnostic and independent of the IT methods being used (i.e. can be applied in any situation where a Business Need is being identified and satisfied) it can be assessed by the business, by the TSP, or jointly. Whereas business and the technology people are mostly separated by language and perspective it is known that the properties of the TBE come from both the business and the TSP people in shared if not equal amounts. With input from both the business and the TSP, the assessment of the TBE becomes a shared activity and, by closing the 'blind spots', a more complete measure of service quality is achieved. The assessment could be carried out by any party independently of any other, if necessary, provided there is visibility of the experience being delivered and it is remembered that service quality can ultimately only be judged by the customer, i.e. the business stakeholder, and governance by the executive or board of directors.
- Governance can be further enhanced by defining the set of interest to the audience, using the navigation tools and then presenting the findings in an interesting and relevant manner. For example, the presenter can compare assessments of services throughout the business hierarchy and show what Alignment Tasks are scheduled, etc. Presentations for management could be compiled as a tour starting with the most important for that audience and finishing with those of least concern, or could be viewed by considering the impact of a Change, or by viewing the degree of closeness (biggest gaps first), or by Key Customer, or Key Supplier, or by TSP. Performance can be reviewed by comparing Agility

post Change. Another presentation or analysis option is to view the dashboard as a time lapse from some point in the past to the present day to view overall improvement or deterioration in alignment, or issues and responses at points in time monitoring the Agility of Response to Change, i.e. time taken to bring about actions to address requirements from the point in time when they were first raised or where a Change is deemed to have occurred. Perhaps during the time lapse it would be of interest to point out the frequency of amendments due to Hindsight. Another analysis could focus on Contagion and the implications for future planning. Finally, the findings on Closeness, Agility, and Contagion could be summarised and used to assess the opportunity for TSP improvement with management tasks drawn from the analysis of the entries. All of this can be achieved whilst being totally confident that the information being presented is completely up to date.

- All stakeholders can be fully engaged in the creation of value though decisions will have to be taken about next steps. This is likely to lead to more business opportunity. All stakeholders are invited to make continuous assessments of service, each has a role in putting matters right, response is quick, feedback is immediate, and the measures of success are clearly defined. This makes it ideal for organisations that not only operate in a dynamic market place but those that have a collaborative culture, or believe that they need to develop a collaborative culture, and who wish to enable business agility. Needs arising from the value ideas collected will be assessed against longer term plans and Alignment Tasks will need to be created and funded. The administration of this will be a Gold Team responsibility requiring the ultimate sanction from the executive or board of directors in their governance role.
- Managing the TBE provides an acid test for determining how well technology is supporting the business, and has the confidence of the business. This makes Active Management and Governance an ideal solution where AI is playing a significant role in the business to ensure that the automated functions are sensibly reflecting the needs of the business and properly adapting to new business circumstances. This might include delivery methods which may themselves eventually be automated.
- This is an effective management and governance solution for business technology even in complex enterprises when there are multiple businesses and multiple TSPs. It is always business-focused, i.e. governance is restored to being about the how the technology is supporting the business from the viewpoint of the business.
- If IT management was sometimes seen to be weak because of the blind spots caused by the methods used, leadership was sometimes lacking as well. By exploring the granularity of the data as described above the governance process can undertake to explore the leadership and budgetary provision associated with the business technology elements in detail and re-align them as appropriate.

Thus, governance is transformed from being an occasional boardroom agenda item to becoming a key control point for the full scope of the business and technology relationship. Whilst there will be periodic reviews of long-term strategy to fix annual budgets that may require a committee structure it is likely that a more dynamic management process based on the continuous assessment and the active management of the business experience will help the business and the TSP to maintain better alignment, be more agile and ensure improved success rates. The stakeholder is after all the final arbiter of what is needed. This allows organisations to deal with a broader spectrum of computing issues and connect technology and the business more effectively and on more levels as is called for in the academic literature.

Active governance then is the continuous process of dynamically responding to these challenges by allocating resources in order to manage the short-term and long-term demands of the business according to cost, benefits, risk, impact, and time to deliver. The purpose of the active management framework is to both improve Closeness and Agility without being too constrained by technologies or methodologies.

What Is Missing

What we know is what we are told, or what is revealed by the insights that Active Management and Governance provides. What we don't know is probably what we are most concerned about. Chapter 5, 'The Automated Future', is about what else Active Management can tell us from the other clues provided by the text and other aspects of the data that we are collecting from our stakeholders. It's about how automation can maybe help us to know more than we have been told and so do more than is expected.

References

Harari, Y. N. (2011). *Sapiens: A brief history of humankind*. Harper.

Nemeth, C. (2018). *No!: The power of disagreement in a world that wants to get along*. Atlantic Books.

Sayed, M. (2019). *Rebel ideas - The power of diverse thinking*. John Murray.

The Automated Future

5

In absolute terms the endgame and the possible routes forward are uncertain and so the first section of this chapter considers the general relevance of the BTRM using a number of different future scenarios. The following sections then consider what can be done in the shorter term and six pathways of opportunity are considered for a business wishing to exploit learning algorithms and other non-conscious intelligence to enhance Active Management and Governance. A section is devoted to each of these. Section two looks at the validation of the technology supply chains, section three, detecting patterns, section four, language and interpretation, section five, ethics and bias, section six, misunderstandings, inconsistency, and misrepresentation, and section seven, 'Very Agile'.

Looking at the longer term, we can expect that automation will be playing a much greater role and we know that Active Management and Governance is based upon a model, the BTRM, that also lends itself to automation. If everything is eventually automated where does the specification come from and how are such system solutions controlled for the good of the business? For society? Sections eight and nine of this chapter focus on enhancing automation and controlling automation.

The final section 'When the business is technology' serves as an introduction to Chap. 6, 'The BTRM as a Universal Management Framework'.

Future Relevance of the BTRM

Active Management and Governance facilitates the collection of information from potentially all stakeholders, wherever a stakeholder is physically located, undertaking whatever kind of work, and in whatever business context. This is important information about the business and how it is using business technology in a form that is useful to the business. The management processes have been described earlier and explain the democratisation already built in to the assessment process, the governance processes take care of what is good for the business, and, the coding system already contains prompts that would allow stakeholders to comment on societal

D. Miller, *Managing Agile Business Technology*, Management for Professionals,
https://doi.org/10.1007/978-3-030-90598-9_5

impact if the business chooses to give this weight. This is Big Data being applied to the management and governance of business technology. There is nothing new about Big Data but it has not been applied to this problem area before in this way, but it will.

New technologies will need to be considered but their importance to us will be assessed by their value-in-use, i.e. what they do for us. Thus, we can try to describe the differences between a new technology and the computing of the last 50 years but the value will be in the resulting algorithms, other outputs, and what it helps us to achieve in business terms. Thus, although the underlying technology may change, the relationship between business and technology remains constant, i.e. the BTRM still holds true.

In consideration of each of the interconnected elements of the Active Management and Governance dashboard:

- The Change concept doesn't change though the nature of Changes reported may.
- The Business Need concept doesn't change.
- The TBE and its dimensions are generic and don't change though over time there may be new attributes that need to be considered to reflect the underlying technology, how it must be managed, and what it is able to achieve, etc., but in the meantime the flexibility of the textual explanations will enable any experience to be captured.
- Service Requirements are generic and will still exist broadly as today even though the underlying technology will change.
- Alignment Tasks will still be undertaken though some may be initiated automatically.
- The Key Stakeholder Groups graphic will still identify and be organised around key processes and, even if staffed entirely by robots, each will still have an owner. Thus, from an Active Management point of view they will still function as they are conceived to function right now though there may be some consolidation of groups and some new groups to reflect the underlying technology.
- The TBE Radar graphic doesn't change unless the TBE dimensions change.
- The Causal Chain graphic doesn't change because the logic still applies and navigation will still be necessary.
- The Impact on Supply graphic: no changes to any of the service types, i.e. to Service Management, Business Planning, Application/Digital Services, Transition Services, or Infrastructure Services though the resource implications will reflect the underlying technologies.

Active Management and Governance has been tested against various future scenarios. In Table 5.1, below, the first four scenarios (numbered 2.1 to 2.4 and set in bold) were suggested by others (Gray & Hovav, 2007) and used as a test within the research model. The fifth scenario (2.5 The eventual) is where we are likely to end up—the only issue is when, and via what route. What this analysis shows is that the Active Management and Governance system is likely to be important in any future foreseeable eventuality.

Table 5.1 Future possible scenarios

No.	Scenario	Current IT delivery model observations	The importance of Active Management and Governance
1	**Back to the dark ages: Recession and monetary crises, energy crisis, globalisation slows, unreliable communications, return to centralised IT, aging infrastructure, little innovation, Moore's law ends, IT support requires more staff but people not attracted to the profession.**	Current situation continues. Success rates associated with IT investment likely to remain unchanged at 35%.	Offers a clear improvement opportunity to provide services that are closer to be needs of the business. Important if there is a desire to reduce costs and risks.
2	**Slow growth: Information systems evolve slowly, innovation not at the pace of the last 25 years, Moore's law slows down, poor telecoms infrastructure and internet speed very slow to improve, transmission of corporate data not trusted. Still islands of IT in companies. Reduced requirement for IT staff.**	Current IT delivery model sustained by islands of IT but as Moore's law slows businesses become more concerned about success rate of 35%	Offers clear improvement opportunity over current situation as businesses look for better value as well as better quality.
3	**Technology driven: Highly developed telecoms infrastructure and highly trained (technically) workforce. Internet becomes segmented and new computing devices emerge. Large enterprises outsource most of their computing relying on standard packages and automated software coding to ensure quality. Cybercrime is low. Focus on development of automated processes. Lower demand for IT staff (40% of 2006 level).**	Success rates may still not be good enough. Closeness of alignment and agility of response will be most important.	Essential if businesses are to become 'Very Agile'.

(continued)

Table 5.1 (continued)

No.	Scenario	Current IT delivery model observations	The importance of Active Management and Governance
	Alignment expected to be low.		
4	**Nirvana: Unlimited low-cost ubiquitous internet. Secure. Computing as a service is universal. Few IT departments. Competitive advantage gained from way in which they combine modules to achieve alignment. No problems with information security or cybercrime. Requirement for IT staff confined to central planning, high-end consultants, and low-end technicians.**	Probably only relevant to developers and data centre managers and then not good enough.	Essential: businesses will only be able to judge quality according to their experiences of using the technology.
5	The eventual: A self-adaptive, self-managing, world of cognitive informatics: Organisations will respond to constant change and alignment will be assessed by key stakeholders according to closeness and agility criteria which will essentially be the same in concept as today. Biggest threat will probably be from state-initiated cyber warfare. Other changes perhaps from increasing number of pandemics combined with less travel and less use of public transport.	The current delivery model will be irrelevant in this context except for manual interventions and then will not be trusted and would expect to have to be controlled within the context of a wider management framework.	More work will be undertaken by machine. Key workers, customers, and suppliers will assess service quality using the principles of Active Management and the BTRM. Technology will be able to respond to the needs of the business. Controls must be highly developed to prevent rogue actors masquerading as stakeholders, devices or services to create false entries and unwanted outcomes.

The future will involve more automation and Active Management and Governance will be no exception. These technologies will serve to further increase our understanding of the relationships defined by the BTRM, the world in which we operate, and further enhance the management and governance of business technology.

We can speculate about the eventual automation endgame and the technologies that might be used but this would probably not be accurate. However, while ever there are people performing tasks in Key Stakeholder Groups with changing business needs requiring technological solutions it seems that it will be appropriate to manage their Total Business Experience. The Closeness of Alignment and Agility of Response will remain key measures of success.

Validation of the Technology Supply Chains

All assessments of technology products or services are important. If improvements are to be made, we must correctly distinguish between the service providers when commenting about products and services in order to identify their source and who should be approached to address the issues being raised. For the purpose of Active Management and Governance the conventions for identifying the TSP as used in the text, and described in the glossary, will in the most part need no further explanation and will be perfectly satisfactory. Identifying the correct supplier can, however, sometimes be more difficult where there are multiple TSPs.

The correct identification will depend upon the perspective of the stakeholder and the supply arrangements for particular services within that business or within that stakeholder's KSG. In some cases, the identification of the supplier may be open to interpretation. In complex supply situations, the service supplied (and supplier) can be identified from the business, the stakeholder KSG, the service, the textual explanation of the issue and what should be done, or by supplier name, or by some or all of these. Some errors can be corrected easily but if there is some inconsistency in the reporting then this may necessitate a member of the Gold Team contacting the stakeholder to check the facts. Alternatively, an algorithm may learn to either suggest a correction or correct the entry for us from the information supplied. AI can help where there are many suppliers or where a stakeholder gets this wrong, e.g. where the stated supplier doesn't match the service and stakeholder combination for that KSG. As the supplier recognition improves it may be possible for the stakeholder not to have to be concerned about this issue leaving it entirely to the AI to determine.

Detecting Patterns

Patterns can be detected manually but automation is likely to be more successful. Five examples are considered here for such automation but in practice there are likely to be many more.

When stakeholders make assessments, it can be argued that there is some subjectivity in the process and the measure of how close the aspect of service that is being assessed is to the ideal can be challenged. In practice this is more of a hypothetical argument than might at first be imagined, e.g. no scores determined in a consultancy situation have ever been disputed. However, by using AI techniques it is possible to learn about how individuals or groups differ in their assessments and so better understand when situations have become abnormal or even critical. Whilst very few people ever award 100%, some will consider 75% to be a high score whilst others may consider that to be more of an average score. Collaboration, triangulation, or consensus, together with discussion about the reasons for any shortfall, and disclosure of opportunities generally bring scores in a given KSG closer together. Also there tends to be a wider range of scores generally across the service spectrum as stakeholders become more analytical in their assessment because reasons have to be given for their ratings. Understanding the dynamics of individual behaviour related to the scoring of services is helpful in setting alerts when assessments fall below a particular threshold, i.e. the system could set group alerts based upon what it has learnt about how people make their assessments rather than using some arbitrary number.

At senior levels in the enterprise, there is a need to know if all KSGs and divisions are adequately supported by services. The system will enable support to be compared but it would be helpful if AI could assist by making comparisons and alerting the business to variations across the enterprise that need further analysis. Examples might include inequalities across KSGs within a division or between divisions, best and worst cases, best and worst vs average, etc., so that the reasons can be explored and addressed. The TBE Radar Diagram on the dashboard could be used to show the multiple plots that are being compared or contrasted when these conditions are detected.

The research showed that Change was often a precursor to new Business Needs, reduced TBEs, and an early warning that technology services may be needed, and yet Change frequently went unreported. In many ways this is the fault of past methods where we 'trained' business people not to speak to us, i.e. the IT function, until they fully understood their detailed requirements. Sadly, as a consequence, businesses often see IT and the TSP as a problem to be overcome leading to serious frustration across the business and a serious lack of knowledge around the impact of Change. As a result of using Active Management and Governance we can expect that Change is being explicitly reported more often and that other categories of data are being added to construct Causal Chains resulting from the Change. Making Change visible allows everyone to assess the impact on their own role and to see the impact on others but by using AI it may be possible to fill in more of these gaps, e.g. where a Change hasn't been reported, where a link to a Change may be implicitly assumed but it could be made explicit, or where a Change has been reported but where no consequential chain of entries exists where they would have been expected, e.g. in the case of Contagion.

If there is a pattern to Contagion, then the enterprise would probably benefit from knowing about this sooner rather than later, i.e. by alerting the Gold Team perhaps to certain assessments that may be indicators of something that could spread rapidly throughout part of the business. In particular where Change is implicit rather than explicit, the enterprise can perhaps predict Contagion and so react at the first sign of what might be the start of a pattern. It may be that there are several patterns here that each require different responses and so, by using AI to assist us with the disaggregation, we might detect more of them.

Closeness of Alignment and Agility of Response are key governance measures that make sense to business people. Thanks to the research, both of these can now be defined and managing them is usually easy. In a less stable environment, where multiple Changes are taking place and management is responding to many new Business Needs, or in large and/or complex enterprises, tracking the effectiveness of individual Alignment Tasks will be more complex. In such circumstances AI could be used to separate these and so simplify this activity.

More About Language and Interpretation

The BTRM and its ontology are the catalysts for changing the way we work. The power is not just in the model, or the categories of data and the relationships between them, or in the classification of and scoring of experiences but in the rich textual explanations provided by stakeholders and the contextual narratives created by the Causal Chains. Whereas coding and classifications are limiting, natural language allows unlimited explanation. AI can help by unlocking the full potential from the patterns of expression within these rich textual explanations. Not all business expressions, technical terminology, natural languages, and certainly not all local expressions, will be universally understood across the extended enterprise without translation, and may require cognitive interpretation. AI has several roles to play here so that not only the actual words used by the stakeholder, but also the meaning can be understood universally through instant translation and interpretation. In some cases, the words actually used by a stakeholder will reveal as much about their personality as they will about their requirements, e.g. historically computers have not handled irony or humour well.

AI will be able to learn how different stakeholders react to Change, to new Business Needs, or to performance issues and will be able to compensate for these differences. This ensures that the importance of such nuances of expression is not lost in translation and similar things can be grouped. For example, several Business Needs may have been entered by different stakeholders and described in different ways but AI may detect that there is much commonality, i.e. determining where they are talking about the same thing and describing the same issue and where a new requirement is being described. The system may raise an alert so that these can be checked with the stakeholders. This could result in the Causal Chains being amended or simply linking the entries determined in this way to have commonality so that they can be addressed using the same Alignment Task.

A key element of creating Alignments Tasks is in grouping similar problem issues and in doing so providing an expression of the value of the Alignment Task. AI could be used as an aid to this process. Explanations and opportunities entered by the stakeholders describing the need will overlap with other entries, and any value add can be apportioned from the proposed solution for consideration and verification. This can be correlated with the scores post implementation for the Sales/

Value Add dimension of TBE for those stakeholders/KSGs to provide a further control.

Used in these ways, AI can thus derive more information from the rich textual explanations submitted, enrich our understanding of the whole of the business and technology relationship, better enable the comparison of assessments by many different stakeholder communities, and more effectively inform Contagion.

Ethics, Bias, and the Business

Ethics in this context is defined as a moral philosophy practiced by stakeholders or groups of stakeholders who are concerned with what is good for the business and its stakeholder communities as a whole. Because machine intelligence lacks empathy automated management systems of all types can attract criticism on ethical grounds because of their capacity to impact people. A business may define its own set of ethics that govern how it operates, how business decisions are made, and how customers, shareholders, and clients are treated. Active Management and Governance is a transparent process for the universal management and governance of technology that is easy to understand and makes sense to business people and which, by its nature, will always reflect those ethics. In its execution what can be entered is defined by the science, undertaken by real people, and decisions based on the entries are sanctioned by real people and everything is visible to other decision makers across the business. The results are also assessed by stakeholders and if these are deemed to have been executed unethically then the experience will be recorded accordingly thus the system is to a large extent self-managing.

There is no data field for 'Ethics' because the concern could potentially arise in any aspect of the TBE. AI, though, could be used to build in more company specific rules to the system or detect ethical concerns within text. It would be important to ensure that in so doing transparency and ease of understanding were not compromised. If successful, it could be used to continuously monitor results or activities and where these are deemed to be unethical, or where there may be issues, to raise alerts.

Bias is a tendency to have opinions that lean in a certain direction, often to the exclusion of clear evidence to the contrary or the opinions of others. Bias and unconscious bias tend to be targeted at people or groups based on their ethnicity, social class, or religion. A cognitive bias is where individuals create their own subjective reality from their bias which then dictates their behaviour. Bias has practical implications for many areas including clinical judgement, entrepreneurship, finance, and management, and can also lead to prejudice, discrimination, and loss of business. Active Management and Governance minimises the likelihood of bias because opinions are visible and actions are auditable.

Thus, bias if present will be visible but AI could be used to detect possible occurrences and gather and report the evidence. For example, the system could report instances where the views of an individual were routinely ignored when that person was continually proved to be correct. Or the system could look for consistently negative opinion being expressed about the suggestions of others to establish whether there was a pattern that could be caused by bias.

Misunderstandings, Inconsistency, and Misrepresentation

Stakeholders are encouraged to record their views promptly without waiting until they have a detailed requirements specification of their needs (because sometimes in an ambiguous world that stage is never reached). For this reason, not all stakeholder entries will be totally correct or accurate but it is important that they don't mislead.

Mostly stakeholders will see the same things but interpret them differently according to how it affects them within the context of the role they perform. Occasionally, as has already been recognised, stakeholders will misunderstand what is happening in which case, with Hindsight, this can be corrected (see Chap. 4, 'Hindsight').

Sometimes, if an inconsistency is discovered, a learning algorithm could detect a possible misunderstanding and challenge the stakeholder to investigate and possibly even suggest a correction.

There is also the possibility that the truth is being deliberately misrepresented in order that a stakeholder may bring about a particular outcome. Algorithms could be developed to detect persistent behaviour of this kind and, if proven, it would be grounds for disciplinary action.

All entries are attributable to a known person in a defined KSG; anonymous or anonymised entries are not permitted so all entries can be followed up thus minimising the possibility of 'fake news', misrepresentation, or deceit, within the data. Any views can be expressed though there is no guarantee that they will all be acted upon. Views are accompanied by an explanation, or 'justification', and so there is an expectation that this will encourage an open and positive behaviour to prevail. Even where people are open about their views but disagree there should be an expectation of collaboration and cooperation whereby a better outcome can be achieved for the enterprise—or even the people within it.

Seeking contributions from external stakeholders, i.e. stakeholders outside of the enterprise, where there is thought to be no commercial risk can be especially insightful and is to be encouraged because the culture and perspective will be different. Where there is a perceived commercial risk, instead of choosing to restrict the access of these external stakeholders, the business may use AI to limit what they are able to view, even validating in some way what they enter to ensure that it is not misleading or detrimental to the well-being of the business. For example, the system could look to see if the views of an external stakeholder were consistently misleading and, if so, restrict access to that stakeholder, limit what that external stakeholder may see, or restrict access by others to that external stakeholder's entries.

'Very Agile'

The first principle of the 'Agile Manifesto' stated that 'The highest priority is to satisfy the customer through early, and continuous, delivery of valuable software'. This has been aided so far by the availability of increasingly agile technology, i.e. technology that is easy to deploy, and agile development processes. The management process, however, is not agile and remains flawed in the areas described earlier. When we change the way we work, we have the opportunity to become 'Very Agile'.

Traditionally, responses have always been after the fact, i.e. when the need has arisen and the detailed requirement has been specified a solution is developed and deployed. It may be that patterns dictate that some responses can be determined automatically, or predetermined, and so implemented sooner.

Automatic responses would be possible in situations where the information required to make the response could unequivocally be determined from entries by stakeholders. This would maybe apply in relatively few cases to begin with but could increase as detection techniques become more sophisticated. Strictly speaking stakeholders are defining the solution here not AI, and investment will be sanctioned by real people.

Perhaps the greatest opportunity is where a Change or Business Need can be anticipated or predicted to take place with some certainty at some unknown point in time in the future enabling a response to be predetermined so that it can be deployed the instant it is needed.

Predetermined responses are enabled by a business preference for small incremental changes to realignment for speed of implementation and validation rather than major upheaval, e.g. as enabled by the adoption of some kind of SOA. They are further enabled by the disaggregation of need. By disaggregating the need it is possible to prioritise what is known will be needed with some certainty as information is entered. Thus, the critically important elements of a response can be prepared in advance for eventual deployment before the full picture is known. This also calls for a more nuanced approach to risk assessment but for a very small outlay at any one time a business could have several important responses ready and tested for immediate deployment in the event of certain foreseen circumstances taking place.

This ability to anticipate or predict how best to prepare the business and what may be important to the business in the future is further aided by Contagion (understanding how the impact of Change spreads through a business by knowing which parts are likely to be affected first) and Hindsight (learning from past events).

The future adoption of other 'Very Agile' features that will further streamline the management and governance of business technology using Active Management and Governance can be anticipated. In particular there is a need to speed up the preliminary discussion stages associated with investment decisions and governance processes. Here, voice recognition will perhaps be the first to be used.

Entering text, whilst providing an infinitely flexible way of communicating, can be time consuming. Verbal communication is much faster but we sometimes leave things out. Voice recognition has developed to the point where intelligent assistants can sometimes be very effective in limited applications but they exemplify the problem of what happens if they are given incomplete information. Now imagine an Active Management and Governance system where information can be entered verbally and the system can respond using speech to ask about links and other possible entries that might be relevant to make sure it's complete, i.e. where there is a meaningful and intelligent verbal exchange that is able to discuss what needs to be entered. The verbal exchange would help to verify the content and the eye would see the resultant natural language text entry that had been created. This coordination of verbal and textual communication would be very powerful. It would allow the system to challenge the user by making them aware of what others had entered and allow people to participate who were carrying out work that involved using their hands and which as a consequence were not available for typing until some later moment in time. Data entry would take place sooner, be much faster, would maybe be more accurate, and could incorporate immediate language translation and other help facilities. It is envisaged that the commit process would be after the stakeholder had seen the textual interpretation.

Similarly retrieving information could be aided by verbal dialogue. A stakeholder at a meeting could request some up-to-date information, or some kind of analysis as described earlier, to be presented to the meeting for discussion by those attending. The system may also offer advice on what might be shown next based on the information given or that has been entered, e.g. to illustrate an opportunity that could be exploited that is relevant to the meeting, or a situation that has developed. Thus, voice recognition plays well to 'Very Agile' but can also help to create value.

Enhancing Automation

The extent to which automation features in the management and governance of business technology is growing. Automation has tended to be introduced for low level routine tasks but the latest types of non-conscious intelligence are said to be capable of out-performing humans undertaking high end tasks.

Further automation based on the ability to anticipate or predict events and needs will eventually give way to truly self-adaptive and self-managing systems given the heuristic, deterministic, and control characteristics of the BTRM. This will continue to be driven by the need to manage the Total Business Experience. The Closeness of Alignment and Agility of Response will still be the key governance measures. This is likely to begin in organisations where the possibilities are limited but eventually can be expected to operate even in very complex and large-scale contexts.

The characteristics of automation can also be applied to improving Active Management and Governance itself. The BTRM by its definition and through its utilisation of textual explanations can fully describe the relationship between the business and its service providers (see the Appendix for the full ontology). Systems based on the BTRM learn more about the relationship as opinion is given through the stakeholder entries and it is possible for this heuristic property to maintain the attributes (dimensions and properties) of the categories of the BTRM so that they are kept updated with anything new and of significance where this is generic and can be considered to be a legitimate aspect of the Total Business Experience. This will be achieved by learning from the rich textual explanations provided within each entry. Using AI capabilities, Active Management and Governance would learn about possible new dimensions and properties and report these as potential system changes.

The BTRM is deterministic, i.e. it is capable of producing the same output from a given set of conditions. We have seen also that it is seen to be capable of indicating that a Change that has not been reported has occurred, determining which parts of the business are affected, what needs to be done, and how these impact on supply. By considering Contagion, it can also understand how this is likely to develop and spread through the enterprise. Because stakeholders describe their circumstances and needs in different ways (and sometimes in different languages) AI could be used to compensate for these differences to ensure that like things are grouped together (see above, 'More about language and interpretation') or disaggregated so that they can be treated separately. AI could also be used to detect, from the use of language, how stakeholders across and within organisational hierarchies are thinking about systems and whether they are likely to be inclined to invest further in realignment possibilities.

The control characteristics of the BTRM are Closeness of Alignment and Agility of Response. Statistics for these are continuously available and are of universal appeal answering such questions as:

(a) Are we aligned?
(b) How do we achieve alignment?
(c) When was the alignment task deployed?
(d) Was the alignment successful?
(e) How long did it take us to achieve alignment after the first indication of the issue, i.e. what was our Agility of Response?
(f) How can we improve on this the next time around?
(g) With Hindsight, and looking back at the original entries, does anything need to be corrected to improve the response the next time these circumstances occur?

AI can also be used to notify stakeholders when they will be affected by an Alignment Task being completed, i.e. not just those stakeholders with entries linked to the Alignment Task. By prompting these stakeholders for re-assessments post completion, and tracking those responses, feedback is maximised.

Controlling Automation

When self-adaptive and self-managing systems are routinely being used in business, we have to ask how we, as humans, stay in control of them.

There is no doubt that some processes will be improved by using such non-conscious tools and in limited ways these may be superior to human consciousness, e.g. by automatically backing out an Alignment Task where many postdeployment problems are being reported. Most shareholders, though, would probably be reassured to know that the management and governance of business technology will be controlled using conscious (human) intelligence and real people will be overseeing most decisions at all levels of the enterprise. Management has to deal with the known and unknown. At the pre-analysis and pre-experience early stage of a Change where there may be many assumptions there may be more unknowns than knowns. At these early stages systems maybe haven't yet experienced the Change and so do not know how to react; only humans can know or speculate about what the system does not yet know. In this context AI will assist, but probably as augmented intelligence rather than wholly artificial intelligence. Process and investment in such situations will always be sanctioned by real people, in fact real people will continue to have important roles to play throughout the enterprise.

Automation has come a long way and we have by no means reached the limits yet of what is possible. However, when people hypothesise about the future extent of automation, I suspect it is often over stated and over worked. Non-conscious intelligence will be pretty dumb when experiencing something for the first time because it will have no reference points whereas humans are more resourceful and more adaptive to change than machines. This means that in the complex, ambiguous, dynamic world characterised by turbulent markets, shorter product lifecycles, disruptive technologies, industry structure change, high staff turnover and changes in employment status, increased home working and changes to other working practices brought about by pandemics or the possibility thereof, and the emergence of new types of non-conscious intelligence, etc., the BTRM is still valid and it is important that humans retain control.

In large and complex organisations, the following are always likely to be the responsibility of humans:

- The executive control of the business direction and Active Management and Governance including investments in business technology, budget, and leadership decisions.
- The business of each KSG ('an organisational grouping associated with a discreet business mission or group of activities and hence having a unique role in the extended enterprise'), or group of KSGs, and the assessment of their outputs.
- Our suppliers, customers, and consumers of end product or service.
- Alignment and Governance (the key control point for the business in the context of business technology) where Closeness of Alignment and Agility of Response are considered along with cost, impact on the business, risk, value, etc.

Thus, where there are benefits for the enterprise and its wider society, the enterprise long-term is likely to prosper from the greater use of automation but there will be enough real human stakeholders around in the future to ensure that the BTRM is still a valid control mechanism, i.e. we will not be seeing the larger and more complex businesses just staffed by mindless algorithms. Non-conscious intelligence will function better than humans in limited areas of operation and those humans that stay in the game will be those that are able to keep learning throughout their working lives, those especially adept at being able to understand the technology used as well as the markets and processes associated with the core business, and those flexible enough to be able to adapt easily and quickly to the latest challenges. Whilst some technical specialists will still be needed in a business, most business managers will have to be tech-savvy. Roles will inevitably merge and an appreciation of business technology will be especially important at the executive level.

When the Business Is Technology

Where businesses are totally dependent upon technology it's maybe no longer appropriate to wholly delegate or abdicate the responsibility for business technology to the TSP. Business needs to not just retain control but get involved and a method for doing this has been described.

Chapter 6, The BTRM as a Universal Business Management Framework, considers the management philosophy of Active Management and Governance and its significance in business more generally.

Reference

Gray, P., & Hovav, A. Z. (2007). The IS organization of the future: four scenarios for 2020. *Information Systems Management, 24*, 113–120. https://doi.org/10.1080/10580530701220967

6 The BTRM as a Universal Business Management Framework

This chapter considers the management philosophy of the BTRM, and Active Management and Governance as a conceptual implementation, and their potential not just for managing business technology but as a universal business management framework which is consistent with corporate governance practices.

Technology has changed the world and will continue to do so. Technologies are converging, the whole world is connected, and consumerisation and commoditisation make the technology available to everyone, affecting everything we do. Changes seem to happen more quickly affecting more businesses and more people. A small change to the technology can have a big impact on business. It is becoming more difficult to separate the technology from the business. Business people are becoming more technology savvy if not totally technology literate. We are told that future self-adaptive or self-managing systems will constantly adapt to new business requirements but that here the complexity makes defining and testing the requirement difficult if not impossible except at run time (Kephart & Chess, 2003). 'Very Agile' techniques described here contradict this view but one way or another the business will become more automated and make greater use of automated systems as they become available. So what does this say about how we manage the business?

Many businesses even today are totally dependent upon technology in almost every business process. Quite simply, technology is the way we do business. By using Active Management and Governance and so actively managing the business technology those businesses will be effectively managing the business itself. Most would think that perhaps this is the wrong way around and so section 'The Benefits' considers its general suitability for managing business—not just the technology, section 'The BTRM as an Integral Part of Corporate Governance' shows how the BTRM has much in common with corporate governance best practice, section 'Hitting Reset to Change the Way We Work' tries to summarise and conclude what has been described and shows that the thinking behind Active Management and Governance has many similarities with some of the best recent management thinking in other sectors.

D. Miller, *Managing Agile Business Technology*, Management for Professionals,
https://doi.org/10.1007/978-3-030-90598-9_6

The Benefits

In the past the concepts of service and the customer experience, and in the digital world customer experience (CX) and employee experience (EX), have been recognised empirically but never fully explored, never fully defined, never been part of the delivery model, and never incorporated into a control system—even though in many organisations almost every business process is now supported to a greater or lesser extent by technology of some kind. Just as IT borrowed from factory processes in the past for its underlying methods (i.e. the PDCA), it might now be sensible for business to borrow from business technology. Just as we need to change the way we manage business technology, as businesses become more dependent on that technology the same principles will allow us to manage the business itself in a more agile manner. The BTRM, or more accurately Active Management and Governance, can achieve for the business what it sets out to achieve for business technology and so it's possible to think of it as a universal business management framework that manages both. The following characteristics are worthy of note:

- It provides the earliest possible warning of changed circumstances facilitating maximum collaboration and an agile response.
- The model is provably correct and makes sense to business people: at its simplest the model checks for new Business Need and tests that outcomes are aligned. Agility is measured from the point where Change affects the assessment to where the response to the Change is implemented. This has given rise to a potentially disruptive management model which is a counter to disruptive technologies and free of artificial constructs. Disruptive technologies dramatically affect a business and the business should not have to depend upon old IT methods for the overall management of its business technology.
- It is suitable for today's complex, multi-dimensional, dynamic, and sometimes dispersed world and for a 'Very Agile' future world, e.g. of anticipated or predicted responses and self-adaptive/self-managing business technologies.
- It manages the big picture providing transparency, control, and better business outcomes: it is possible to understand the full spectrum of the requirement from different perspectives and control the efficacy of the enterprise, technology investment, the IoT, and the many service providers.
- Provides a management capability suited to the real-world context of multiple parallel processes, i.e. where multiple changes are taking place continuously across multiple functions each being affected differently depending upon their purpose.
- Involves all stakeholders in the management and governance of the business and the technology through the early identification and development of opportunities, linking all investment to Business Needs. Governance is simply defined by closeness (as a measure of alignment) and agility (as a measure of response). The benefits are most likely to occur where cooperation and collaboration between stakeholders across the extended enterprise is either routine or being encouraged.

- Addresses the post pandemic 'normal' where stakeholders may be widely dispersed as opposed to being on corporate sites.
- Addresses multi-national and multi-lingual contexts.
- An intuitive model that makes sense to everybody means that little training or retraining is required so it is more sustainable than present methods.
- Addresses value creation—not just value delivery.
- Fine-tunes leadership and budgetary controls: exposes all leadership and budgetary gaps to a 360-degree scrutiny without the distractions of directed personal criticism.
- Enables the introduction of business-focused technology service contracts: the TBE is a measure of real service quality and is provably correct allowing controls on service contracts to be expressed in a form that makes sense to business people.
- Responds quickly to Change and leverages the Active Management and Governance agenda for business, and business technology, across the enterprise for greater business agility and JIT decisions.
- Helps manage scale and complexity: the extended business enterprise associated with new channels, including those made possible by business technology, means that many businesses are now complex and managing the needs of the many stakeholders in the business community by talking to everyone requires some automation: the BTRM facilitates automation.
- Facilitates automation: perhaps the most exciting difference between other models and the BTRM are its heuristic, deterministic, and control characteristics.
- Provides a control for automation, not just the automation of business technology management, but automation more generally by seeking continuous feedback on the way the business is working with or without technology.
- Remains valid in the context of emerging/future technologies: the importance of service is universal and the generic model establishes the TBE as a measure of service allowing service and the Total Business Experience to be managed—regardless of the technology.
- Maps to other methods providing an over-arching management framework so that any investment in other methods is not wasted.
- As will be seen from the next section, the BTRM has much in common with corporate governance best practice.

It is probably worth making a final point on the use of the traditional IT methods. It's now clear that the old methods should only be used within a broader management framework such as Active Management and Governance. There are some TSPs who will reject the idea of adopting the BTRM. There are many who will want to adapt their old IT methods to what we now know, if only to extend the life of an investment in some or all of the bricolage of IT methods that have been developed over the last fifty years. In reality neither approach makes much sense in the long term. Now, more than ever we need to be thinking about how business technology, and business itself, is changing. The BTRM allows us to reimagine the management and governance processes of the enterprise to enormous benefit. No amount of tinkering with a model conceived in history and which reflects the product and production paradigm that existed in that era will make it suited to today's more complex, unpredictable, and agile real-world situation. If the TSP will not adopt the BTRM, then perhaps the business itself will be motivated to adopt it as its universal business management system to provide that over-arching framework.

Introducing an over-arching management framework, whether for the business as a whole or just for business technology, will provide new levels of management control and new business opportunities. This will be particularly beneficial in large or complex organisations where collaboration is encouraged—either within the business or across the extended enterprise.

Not all enterprises will wish to risk changing the way they work with technology without some thought. Once the over-arching framework is in place, the older methods can be reviewed on a continuous basis to determine what can be retained, what becomes redundant, and what needs to be replaced by something else. In fact, as indicated in the section 'Capturing assessments of the Total Business Experience', assessments could be encouraged to reflect on the extent and speed of the adoption of the new methods and the impact this has had on the old methods.

The BTRM as an Integral Part of Corporate Governance

We have seen from the first chapter that there have been many attempts to describe governance in the context of IT. It is often positioned as how business technology investment decisions get debated and agreed, or about how the board on behalf of the shareholders can be reassured about the effectiveness of the management of business technology. Such IT governance is an occasional IT, i.e. technical, process that has been expressed in the literature as being inadequate.

By contrast, we see that Active Management and Governance is a continuous process, involving stakeholders at all levels of the business, of setting strategy, detecting Change and maintaining Alignment (across all of the dimensions of the business experience) and for ensuring that both business and the business technology are managed effectively, and in a timely manner, in order to address the new circumstances. It is based on delivering a managed business experience by appropriate resourcing and leadership to close gaps as necessary. Not only is this essential for managing the business technology, but as business and technology converge it is essential for managing the business itself. In fact, Active Management and Governance is also consistent with, and could be considered to be an integral part of, corporate governance.

This can best be illustrated by taking the UK Corporate Governance Code (FRC, 2018) as an example of good practice in this field. It says that the board's role is to provide leadership to the company and promote its long-term sustainable success, generating value for shareholders and contributing to wider society. It is the job of the board to establish the company's purpose, values, and strategy and satisfy itself that these and its culture are aligned. All directors must act with integrity, lead by example, and promote the desired culture. It should ensure that the necessary resources are in place for the company to meet its objectives and measure performance against them. The board should also establish a framework of prudent and effective controls, which enable risk to be assessed and managed. In order for the company to meet its responsibilities to shareholders and stakeholders, the board should ensure effective engagement with, and encourage participation from, these parties. The board should ensure that workforce policies and practices are consistent with the company's purpose, values, strategy, and culture and that the workforce should be able to raise any matters of concern. There is a need for accurate, timely, and clear information and to promote a culture of openness and debate. The company must have the policies, processes, information, time, and resources needed in order to function effectively and efficiently. Finally, the board is responsible for the integrity of the financial and narrative company statements and it should establish procedures to manage risk, oversee the internal control framework, and manage the principal risks it is willing to take in order to achieve its goals.

The board cannot perform the above roles unless it works with business technology. By changing the way we work with business technology, i.e. by using Active Management and Governance for the whole business, we have more information, more management choices for managing the business- and corporate governance is built-in rather than added on. This is illustrated in the governance framework set out in Table 6.1, below. This is constructed to illustrate how the existing coding structure for Active Management and Governance, without modification, can be used to implement the recommendations of the UK Corporate Governance Code. As can be seen from this table, the governance of business technology, as called for by researchers, is not only able to involve the business more, to address a broader spectrum of business needs, to connect with business more effectively, and at more levels, and for this to be a continuous process, but it actually becomes an integral and important part of corporate governance.

Whilst everyone can contribute, the responsibilities could have been allocated differently according to company preferences. For clarity the Gold Team activity is located here in Table 6.1 within the Governance Criteria column making it a Board or Director/Executive responsibility as suggested throughout this work. The key control point is as always via the Alignment and Governance category of the BTRM where investment proposals, or Alignment Tasks, are assembled (by the Gold Team), investment decisions are made (by the Board of Directors/Executive), and the overall effectiveness is observed by potentially everyone and monitored by the Gold Team). Thus, regardless of whether the preferred technology management model is inclined towards the Executive or the TSP, there is information to hand to ensure that the business can effectively carry out its governance responsibilities.

Table 6.1 shows that the complete scope of corporate governance is summarised by the content of the column headed Governance Criteria and how the information flows through Active Management and Governance. Thus, control of the business, and its technology, takes place at a business level by observing real-world events, the organisation's response and the outcomes in a manner that is entirely consistent with corporate governance best practice.

Table 6.1 The BTRM as a governance framework

No	Category	Governance criteria	Business area responsibility	TSP responsibility
1	Change	Changes to the competitive and trading environment reported in a timely manner that might put at risk the alignment with the business strategy and allowing others to prepare.	All stakeholders able to raise any matters of concern.	Consider possible implications, including 'Very Agile' requirements.
2	Business need	Plan for new processes, features, information, etc. needed in order to continue to function effectively and efficiently, e.g. given the changes likely to take place.	All stakeholders able to raise any matters of concern and possible remedies. Report new processes, features, information, etc. needed in order to continue to function effectively and efficiently, e.g. given the changes likely to take place.	Consider possible implications, including 'Very Agile' requirements.
3	Total business experience (assessments)	Monitor services across the whole of the business and manage by prioritising needs. Monitor entry rates to ensure feedback is coming from across the whole of the extended enterprise.	All stakeholders continuously report issues, i.e. shifts in satisfaction levels, as an effective control to enable risk to be assessed and managed and value to be created through remedy suggestions.	Maintain service levels. Ensure leadership and budget is adequate across all TBE dimensions. Make own assessments and collaborate with business stakeholders on possible solutions.
4	Service requirement	Oversee demand and resource management.	All stakeholders able to request resource that is thought to be needed to define requirement/remedy or next step towards this.	Resource estimates, managing resource availability, service delivery.
5	Alignment and governance	Working alongside the gold team, approve gold team outputs including investment proposals taking into account strategy, value propositions, contagion and possible longer- term	Update scores post implementation.	Manage work streams and delivery.

(continued)

Table 6.1 (continued)

No	Category	Governance criteria	Business area responsibility	TSP responsibility
		implications. Gold team: Monitor integrity of data, existence of causal chains and Spectrum of requirements, value opportunities. Create alignment tasks. Report value creation. Continuously review closeness of alignment and agility of response learning where appropriate from hindsight. Respond to alerts.		

Hitting Reset to Change the Way We Work

Industry leaders are once again challenging the way we work. In a 2021 survey of IT leaders, it was noted amongst the responses that a number of people had remarked on how there was a constant need to keep pace with change. One respondent had also suggested that perhaps now was the time for the industry to hit reset on its IT management methods, toolsets, and skills to improve success rates and so enable business to engage better with IT. The door is open, it just needs to be pushed.

The concepts, management philosophy, and a conceptual system (Active Management and Governance) based on the BTRM may be the reset button that people are looking for where alignment and governance are sought and where the unexpected may occur. Together they enable us to better manage Change and our Response to Change, and will eventually enable business people to do the same, perhaps as an integral part of the corporate governance process. This certainly changes the way we work but the content of this book can only scratch the surface of what is possible. The book has focused mainly on the basics and offered some new possibilities in terms of how we analyse and manage real-world landscapes. Now just consider the more social and cultural dimensions of alignment and the opportunities that might arise if these could be better harnessed and scaled up to work across large complex enterprises.

The strategist when given a brief will seek input from as many people as possible using an extensive range of contacts at all levels within the enterprise, and from others in external organisations such as customers and suppliers. The information will be analysed and proposals will be based on what is known at the time. But the strategist is dealing with a multi-dimensional problem with many unknowns that is changing constantly. The scale and complexity of most businesses makes this difficult for all but the most able business analyst. Use too many analysts and the information gathered becomes fragmented. Use too few and the information is going to be incomplete. Regardless of the number of analysts used the strategist will worry about what is not known. Within a short period of time whatever the analyst produces will be out of date. Active Management and Governance enables us to hit that reset button to gather information on an industrial scale—not just occasionally but continuously.

The philosophy encapsulated within the BTRM and Active Management and Governance is consistent with some of the best recent management thinking. Just as Active Management and Governance enables business experience to be managed on an industrial scale, each of the following examples of intellectual insight could also be enabled on an industrial scale:

- Jeffrey Liker in 'The Toyota Way' describes the principles and behaviours that underlie the Toyota Motor Corporation's managerial and production philosophy which is highly regarded and well documented (Liker, 2004). The 'go and see' (*genchigenbutsu*) has been mentioned in the text but there are other Toyota Way principles, e.g. pull systems avoid overspend, the culture of stop to fix problems (right first time), employee empowerment, use visual controls so no problem gets hidden, respect your extended network of partners and suppliers by challenging them and helping them to improve, make decisions by consensus, learn as an organisation through relentless reflection and continuous improvement—*so many similarities to Active Management and Governance.*
- James Surowiecki in the 'Wisdom of Crowds' tells us that large groups are better than a few experts for solving problems (Surowiecki, 2004)—*there may be a time for focus groups but Active Management and Governance monitors and effectively engages with whole/multiple communities.*
- Nassim Nicholas Taleb in 'The Black Swan' when describing the 'impact of the highly improbable' says that what we don't know is far more relevant than what we do know (Taleb, 2010)—*while PDCA is based on what we know, Active Management and Governance is open to what we don't know* and its impact.
- Daniel Kahneman in 'Thinking Fast and Slow' advocates a combination of System 1 (automatic and effortless control) and System 2 (focusing effort on where it is needed) thinking for managing huge complexity (Kahneman, 2011)—*this could be describing Active Management and Governance.*
- Eric Ries in 'The Lean Startup—How Constant Innovation Creates Radically Successful Businesses' extols the virtues of validated learning (Ries, 2011): He says that, as customers don't really know what they want, the 'just do it' thinking is not appropriate but on the other hand it is possible to fall victim to over-analysis. He argues that in both cases it's easy for the entrepreneur and the developer to delude themselves that they are on the right path. The solution, he argues, is validation; perhaps using a minimum viable product (MVP) followed by continuous deployment based on validated learning—*or alignment tasks and continuous assessment as in Active Management and Governance.*

- Anthony King and Ivor Crewe in 'The Blunders of our Governments' explain that blunders are not random but there is a pattern that can only be changed by changing the way in which people function within the system (King & Crewe, 2013)—*the blunders continue so let's change the way we work by implementing Active Management and Governance—both for technology and for the business.*
- Charlan Nemeth (Nemeth, 2018) in 'No!: The Power of Disagreement in a World that Wants to Get Along' suggests that the processes for how we make decisions and judgements are flawed and explains how it is that others influence our judgements: people influence us in a distinctly different manner depending upon whether they are a majority and have consensus or whether they are a minority voice expressing dissent—*but using Active Management and Governance, where everybody has a voice, the likelihood of such bias is minimised.*
- Matthew Sayed in 'Rebel Ideas—the power of diverse thinking' says that cognitive diversity based on differences of perspective, insights, experiences and thinking styles helps to build a more complete picture (Sayed, 2019): complex multi-dimensional problems require cognitive diversity and coverage across the problem space. *Yes, and this is facilitated by Active Management and Governance.*
- Dick Elsy chaired VCUK, the consortium of British engineering and manufacturing companies that were to respond to the UK government's call for ventilators at the start of the Covid-19 pandemic in March 2020 (Elsy, 2021). In his article for The Institution of Engineering and Technology (IET), 'Ventilator teamwork shows what's possible with innovative thinking', he describes its success and lists the learning points as follows: a clear purpose, openness, fit for purpose structures organised around the key parcels of work. 'The whole approach was needs-driven with a commitment to aiming high and failing fast. Although the consortium involved people from different organisations and cultures, the sense of cohesion was huge'. As Dick Elsy says, 'Most of us working on VCUK ask ourselves: why can't we work like this all of the time? What stops us from unleashing this potential in our own organisations?'. *There was no suitable management framework for this at the start, it evolved out of necessity, but clearly Active Management and Governance and the thinking around key Stakeholder Groups would have been highly suited to the task.*

These are often the antithesis of what happens in the world of information technology today. These examples further underpin the importance and potential additional value of Active Management and Governance as a universal business management framework which enables techniques that support these intellectual insights to be allowed into the field of technology management.

Managing Agile Business Technology has been a book that has been honest about the historic problems associated with managing IT, problems that have been identified by academics and technology professionals alike. It has also been about the discovery of the BTRM and how the concepts, management philosophy, and the ideas embraced within Active Management and Governance evolved and how they can enable us to manage business technology in our complex, ambiguous, and dynamic world to achieve business alignment and agility. There has never before been a coherent business technology management and governance model that takes into account of the true nature of our world.

The book has been written to provide organisations with enough information to use it to either simply stimulate management thinking or to engineer any of the solution options as suggested in the introduction. I hope that there will be many successful such implementations. There have been few developments over the last 50 years that have enabled us to change the way we work, but change we must. There have been fewer that have brought business and technology together in the way that this has the potential to so do. Originally conceived as a solution to today's IT management problems, it is simultaneously a model for managing business technology in a 'Very Agile' manner, managing automation, and managing business. If this is the start of the fabled merging of business and technology, then stakeholders will no longer be assessing the TSP but the technology itself: 'plus ça change, plus c'est la même chose', or, the more things change, the more they stay the same....

References

Elsy, D. (2021). Ventilator teamwork shows what's possible with innovative thinking [comment]. *IET Engineering & Technology Newsletter*.

Kahneman, D. (2011). *Thinking, fast and slow*. Farrar, Straus and Giroux.

Kephart, J. O., & Chess, D. M. (2003). *The vision of autonomic computing computer* (pp. 41–50). IEEE Computer Society.

King, A., & Crewe, I. (2013). *The blunders of our governments* (1st ed.). Oneworld Publications.

Liker, J. K. (2004). *The Toyota way*. McGraw-Hill.

Nemeth, C. (2018). *No!: The power of disagreement in a world that wants to get along*. Atlantic Books.

Ries, E. (2011). *The lean startup - how constant innovation creates radically successful businesses*. Penguin Books (USA) ed. Crown Publishing Group.

Sayed, M. (2019). *Rebel ideas - the power of diverse thinking*. John Murray.

Surowiecki, J. (2004). *The wisdom of crowds*. Doubleday.

Taleb, N. N. (2010). *The black swan: Second edition: The impact of the highly improbable*. Random House Publishing Group.

The UK Corporate Governance Code. (2018).

Appendix

The data fields defined in this appendix have been derived from the research and other sources during the writing of this book to enable the reader to add substance to what is being described and to begin to specify the requirement for building an Active Management and Governance solution. Please note that it has not been subjected to any kind of machine or compiler validation and may contain errors or inconsistencies though it is hoped this is not the case.

Table A.1 Data fields for the Header Data required for every entry

Header Data Fields
The Header Data is standard in every entry and identifies the stakeholder submitting the entry.
Entering Stakeholder's Name: The full name of the stakeholder (a pseudonym, or anonymity of any kind, is not permitted).
Entering Stakeholder's KSG: Normally the KSG to which the stakeholder is permanently assigned.
Entering Stakeholder's Job Title: The job title relevant to the entry—normally the permanent job title.
Entering Stakeholder's Role (optional): The role relevant to the specific work activity in the context of the issue prompting the entry.

Table A.2 Data fields for Change

CATEGORY: Change.
An explanation of a Change that has or may take place which represents an opportunity or threat to the business. Further information about the same Change can be added as and when more information becomes available provided it does not invalidate the original entry.
Header Data. Stakeholder name, KSG, etc. See Table A.1.
Date and time stamp. When the entry was made.
Name given to the Change. Name or short phrase used as identification.
Status. Valid (blank), Invalid (X), or In Need of Checking (?). See Hindsight.

D. Miller, *Managing Agile Business Technology*, Management for Professionals,
https://doi.org/10.1007/978-3-030-90598-9

Reason for Status Change. Stakeholder's reason for changing the status.
FURTHER INFORMATION:
Date and Time Stamp. When the entry was made;
Long Description of the Change. Further information about the Change that has become available since the original entry was submitted.

Table A.3 Data fields for Business Need

CATEGORY: Business Need
A statement of need expressed using four dimensions.
Header Data. Stakeholder name, KSG, etc. See Table A.1.
Date and time stamp. When the entry for this Business Need was last updated.
Name given to the Business Need. Name or short phrase used as identification.
Links. Connections to any Change that that is considered to be a Causal Link.
Status. Valid (blank), Invalid (X), or In Need of Checking (?). See Hindsight.
Reason for Status Change. Stakeholder's reason for changing the status.
DIMENSION: Business Context
Date and time stamp. When the entry was made.
Long description. The stakeholder's view of a modified business context that must be considered and expressed in whatever terms make sense to the stakeholder, e.g. the impact of a technology on a value network, or position the need with respect to the past, current, and future states of the business, etc.
DIMENSION: Business Strategy and Objectives
Date and time stamp. When the entry was made.
Long description. The stakeholder's view of how the overriding aims of the business must be modified/has been modified.
DIMENSION: Needs Gap
Date and time stamp. When the entry was made.
Long description. The stakeholder's view of what is needed to achieve the strategy and its objectives or to carry out a role, e.g. a new/revised business process or capability, better data, a new service, or a technology, etc.
DIMENSION: Business Expectation
Date and time stamp. When the entry was made.
Long description. The stakeholder's view of what kind of response is expected from those addressing the Business Need. Is the Needs Gap expected to be fully addressed reasonably quickly? If not should a short-term tactical/partial solution be considered and, if so, what aspect is the most important? Solution options might be suggested even though these might be rejected in favour of others later.

Table A.4 Data fields for Total Business Requirement

CATEGORY: Total Business Experience (TBE)
An assessment of the stakeholder's perception of the suitability of the Technology

Service Provider's (TSP's) products and services at a point in time using 10 dimensions.
Header Data. Stakeholder name, KSG, etc. See Table A.1.
Date and time stamp. When this assessment was entered.
Links. Connects to any Business Need or Change that is considered to be a Causal Link.
Short description. Short phrase to explain the reason for making this assessment of the TBE or point the reader to the dimension most affected where further detail is given.
Overall Total Business Experience Score Out of 100. A rating calculated to express the TBE of an individual which can be aggregated as appropriate to a group, division, or the business as a whole. Expressed as a percentage of the stakeholder's 'ideal' needs, where a score of 100 is 'ideal', and is calculated according to a formula using the Core Services and Service Management scores (below).
Core Services score out of 100. A mean of the ratings for the dimensions associated with delivery services (also see Table 3.1, Chap. 3)
Service Management score out of 100. A mean of the ratings for the dimensions associated with the management and planning of services (also see Table 3.1, Chap. 3).
Status. Valid (blank), Invalid (X), or In Need of Checking (?). See Hindsight.
Reason for Status Change. Stakeholder's reason for changing the status.
DIMENSION: Business Awareness (BAWARE)
How the stakeholder perceives the TSP in terms of its knowledge of the business and its business sector, domain, or environment.
Date and time stamp. When the Business Awareness was last amended, i.e. not the date it was last incorporated unchanged into an overall assessment of the TBE.
Business Awareness score out of 100. The rating is calculated as the arithmetic mean of the latest scores given by the stakeholder for each of the properties below, at the date and time stated above, including those carried forward unchanged from a previous assessment.
Long description. Any comment which helps to explain the assessment calculated for this dimension or point the reader to the property (below) that is primarily responsible where a fuller explanation is available for the change in the score above.
PROPERTIES of Business Awareness:
Knowledge of the Macro Environment (BA_MacroE): The stakeholder's view of the TSP's knowledge of the broader political, economic, sociological, and technical dimensions of the business environment (note that 'Political' also embraces environment/ecological and legal issues).
Date and time stamp. When this property was last amended.
Score out of 100. The stakeholder's latest assessment for this property.
Explanation. The stakeholder's reasons for any assessment that is less than ideal, i.e. less than 100%, perhaps with examples of the difficulties this creates and what should be done.

Knowledge of the Meso Environment (BA_MesoE): The stakeholder's view of the TSP's knowledge of supply and demand within the business's global markets, and their condition and outlook within the context of the macro environment.
Date and time stamp. When this property was last amended.
Score out of 100. The stakeholder's latest assessment of this property.
Explanation. The stakeholder's reasons for the score perhaps with examples of the difficulties this creates and what should be done.
Knowledge of the Micro Environment (BA_MicroE): The stakeholder's view of the TSP's knowledge of product/market segmentation as it relates to the business, the outlook for the business's current and future products/services and knowledge of the key business processes.
Date and time stamp. When this property was last amended.
Score out of 100. The stakeholder's latest assessment of this property.
Explanation. The stakeholder's reasons for the score perhaps with examples of the difficulties this creates and what should be done.
Key Performance Indicators (BA_KPIs): The stakeholder's view of the TSP's knowledge of the key business metrics for decision making and reporting business performance, etc.
Date and time stamp. When this property was last amended.
Score out of 100. The stakeholder's latest assessment of this property.
Explanation. The stakeholder's reasons for any assessment that is less than ideal, i.e. less than 100%, perhaps with examples of the difficulties this creates and what should be done.
Assumptions/SWOT (BA_Assump): The stakeholder's view of the TSP's knowledge of the generally held assumptions about the business including its strengths, weaknesses, opportunities, and threats.
Date and time stamp. When this property was last amended.
Score out of 100. The stakeholder's latest assessment of this property.
Explanation. The stakeholder's reasons for the score perhaps with examples of the difficulties this creates and what should be done.
Business Awareness Budget (BA_Budg): The stakeholder's view of the adequacy of the TSP's budgetary provision for maintaining business awareness.
Date and time stamp. When this property was last amended.
Score out of 100. The stakeholder's latest assessment of this property.
Explanation. The stakeholder's reasons for the score perhaps with examples of the difficulties this creates and what should be done.
Business Awareness Leadership (BA_Lead): The stakeholder's view of the adequacy of the TSP's leadership arrangements for maintaining business awareness.
Date and time stamp. When this property was last amended.
Score out of 100. The stakeholder's latest assessment of this property.
Explanation. The stakeholder's reasons for the score perhaps with examples of the difficulties this creates and what should be done.
DIMENSION: Marketing/Communications (MARCOM)

How the stakeholder perceives the effectiveness of the process by which the technology service provider communicates externally with the business and internally with its own people.

Date and time stamp. When Marketing/Communications was last amended, i.e. not the date it was last incorporated unchanged into an overall assessment of the TBE.

Marketing/Communications score out of 100. The rating is calculated as the arithmetic mean of the latest scores given by the stakeholder for each of the properties below, at the date and time stated above, including those carried forward unchanged from a previous assessment.

Long description. Any comment which helps to explain the assessment calculated for this dimension or point the reader to the property (below) that is primarily responsible where a fuller explanation is available for the change in the score above.

PROPERTIES of Marketing/Communications:

Brand (MC_Brand): The stakeholder's assessment of the TSP's overall brand image or reputation.

Date and time stamp. When this property was last amended.

Score out of 100. The stakeholder's latest assessment of this property.

Explanation. The stakeholder's reasons for the score perhaps with examples of the difficulties this creates and what should be done.

Products, Services and Their Markets (MC_PS&Ms): The stakeholder's assessment of the TSP's range of technology products and services available to the business.

Date and time stamp. When this property was last amended.

Score out of 100. The stakeholder's latest assessment of this property.

Explanation. The stakeholder's reasons for the score perhaps with examples of the difficulties this creates and what should be done.

Communications Strategy and Stakeholder Communication (MC_ExtC): The stakeholder's assessment of the quality of communication from the TSP to business stakeholders.

Date and time stamp. When this property was last amended.

Score out of 100. The stakeholder's latest assessment of this property.

Explanation. The stakeholder's reasons for the score perhaps with examples of the difficulties this creates and what should be done.

Internal Communications (MC_IntC): The stakeholder's assessment of the standard of communication internally within the TSP community.

Date and time stamp. When this property was last amended.

Score out of 100. The stakeholder's latest assessment of this property.

Explanation. The stakeholder's reasons for the score perhaps with examples of the difficulties this creates and what should be done.

Technology Strategy (MC_TStr): The stakeholder's assessment of the approach to technology strategy development for addressing how long-term Business Needs will be met by the TSP.

Date and time stamp. When this property was last amended.

Score out of 100. The stakeholder's latest assessment of this property.
Explanation. The stakeholder's reasons for the score perhaps with examples of the difficulties this creates and what should be done.
Strategic Data Layer (MC_DStr): The stakeholder's assessment of the TSP's architectural interpretation and implementation of what data must be sourced and managed for the business to function effectively (the critical information set) and the current status of this.
Date and time stamp. When this property was last amended.
Score out of 100. The stakeholder's latest assessment of this property.
Explanation. The stakeholder's reasons for the score perhaps with examples of the difficulties this creates and what should be done.
Enterprise Architecture (MC_EA): The stakeholder's assessment of the TSP's interpretation of the enterprise architecture for communicating and validating the blueprint for designing the technology components that support the business.
Date and time stamp. When this property was last amended.
Score out of 100. The stakeholder's latest assessment of this property.
Explanation. The stakeholder's reasons for the score perhaps with examples of the difficulties this creates and what should be done.
Marketing/Communications Budget (MC_Budg): The stakeholder's assessment of the adequacy of the TSP's budgetary provision for marketing/communications activities.
Date and time stamp. When this property was last amended.
Score out of 100. The stakeholder's latest assessment of this property.
Explanation. The stakeholder's reasons for the score perhaps with examples of the difficulties this creates and what should be done.
Marketing/Communications Leadership (MC_Lead): The stakeholder's assessment of the TSP's leadership of marketing/communications activities.
Date and time stamp. When this property was last amended.
Score out of 100. The stakeholder's latest assessment of this property.
Explanation. The stakeholder's reasons for the score perhaps with examples of the difficulties this creates and what should be done.
DIMENSION: Sales/Value Add (VALADD)
How the stakeholder perceives the effectiveness of the TSP in matching its capabilities and solutions to Business Needs.
Date and time stamp. When Sales/Value Add was last amended, i.e. not the date it was last incorporated unchanged into an overall assessment of the TBE.
Sales/Value score out of 100. The rating is calculated as the arithmetic mean of the latest scores given by the stakeholder for each of the properties below, at the date and time stated above, including those carried forward unchanged from a previous assessment.
Long description. Any comment which helps to explain the assessment calculated for this dimension or point the reader to the property (below) that is primarily responsible where a fuller explanation is available for the change in the score above.

PROPERTIES of Sales/Value Add:
Sales Strategy (VA_SaStr): The stakeholder's assessment of the TSP's approach to targeting business and establishing business commitment to technology investment whilst adhering to and supporting the business investment approval processes.
Date and time stamp. When this property was last amended.
Score out of 100. The stakeholder's latest assessment of this property.
Explanation. The stakeholder's reasons for any assessment that is less than ideal, i.e. less than 100%, perhaps with examples of the difficulties this creates and what should be done.
Sales and Account Management (VA_S&Amgt): The stakeholder's assessment of the TSP's more transactional aspects of the sales, pricing, and account management processes. This is frequently the focal point for all formal communication, performance, and financial reporting between the business and the TSP.
Date and time stamp. When this property was last amended.
Score out of 100. The stakeholder's latest assessment of this property.
Explanation. The stakeholder's reasons for the score perhaps with examples of the difficulties this creates and what should be done.
Pre-Sales Support, Innovation, Value Creation (VA_PreSaS): The stakeholder's assessment of the TSP's pre-sales support to the business perhaps by investigating new solutions or services to create changed business models or improved business value.
Date and time stamp. When this property was last amended.
Score out of 100. The stakeholder's latest assessment of this property.
Explanation. The stakeholder's reasons for the score perhaps with examples of the difficulties this creates and what should be done.
Investment Portfolio, Order Book, Sales Pipeline (VA_IP/f): The stakeholder's assessment of the management of the investment portfolio, its content in terms of its relevance to the emerging needs of the business, and its usefulness in a planning context.
Date and time stamp. When this property was last amended.
Score out of 100. The stakeholder's latest assessment of this property.
Explanation. The stakeholder's reasons for the score perhaps with examples of the difficulties this creates and what should be done.
Sales/Value Add Budget (VA_Budg): The stakeholder's assessment of the adequacy of the TSP's budgetary provision for the sales and value add activities.
Date and time stamp. When this property was last amended.
Score out of 100. The stakeholder's latest assessment of this property.
Explanation. The stakeholder's reasons for the score perhaps with examples of the difficulties this creates and what should be done.
Sales/Value Add Leadership (VA_Lead): The stakeholder's assessment of the adequacy of the TSP's leadership arrangements for the sales and value add activities.
Date and time stamp. When this property was last amended.

Score out of 100. The stakeholder's latest assessment of this property.
Explanation. The stakeholder's reasons for the score perhaps with examples of the difficulties this creates and what should be done.
DIMENSION: Finance and Administration (F&ADMIN)
How the stakeholder perceives the effectiveness of the Finance and Administration activities of the TSP, including all statutory, legal, and regulatory requirements.
Date and time stamp. When Finance and Administration was last amended, i.e. not the date it was last incorporated unchanged into an overall assessment of the TBE.
Finance and Administration score out of 100. The rating is calculated as the arithmetic mean of the latest scores given by the stakeholder for each of the properties below, at the date and time stated above, including those carried forward unchanged from a previous assessment.
Long description. Any comment which helps to explain the assessment calculated for this dimension or point the reader to the property (below) that is primarily responsible where a fuller explanation is available for the change in the score above.
PROPERTIES of Finance and Administration:
Contractual Arrangements (FA_Contr): The stakeholder's assessment of the suitability of the TSP's contractual arrangements for work undertaken for the business, e.g. agreements concerning what, when, to what standard, and what payment terms, etc.
Date and time stamp. When this property was last amended.
Score out of 100. The stakeholder's latest assessment of this property.
Explanation. The stakeholder's reasons for the score perhaps with examples of the difficulties this creates and what should be done.
Financial Management (FA_FMgt): The stakeholder's assessment of the TSP's financial control and management accounting activities.
Date and time stamp. When this property was last amended.
Score out of 100. The stakeholder's latest assessment of this property.
Explanation. The stakeholder's reasons for the score perhaps with examples of the difficulties this creates and what should be done.
Compliance (FA_Compl): The stakeholder's assessment of the TSP's processes policies and procedures designed to ensure that the requirements of all legislation, licensing and other mandatory controls are always met.
Date and time stamp. When this property was last amended.
Score out of 100. The stakeholder's latest assessment of this property.
Explanation. The stakeholder's reasons for the score perhaps with examples of the difficulties this creates and what should be done.
Admin Management (FA_AMgt): The stakeholder's assessment of the TSP's management of its administrative activities and controls. Typically includes diary management, travel, hotels, booking training courses, conferencing/teleconferencing facilities, etc.
Date and time stamp. When this property was last amended.

Score out of 100. The stakeholder's latest assessment of this property.
Explanation. The stakeholder's reasons for the score perhaps with examples of the difficulties this creates and what should be done.
Growth (FA_Grow): The stakeholder's assessment of the manner in which the TSP manages its growth and downsizing in association with commercial and financial planning and control. Weaknesses can threaten the sustainability of the organisation.
Date and time stamp. When this property was last amended.
Score out of 100. The stakeholder's latest assessment of this property.
Explanation. The stakeholder's reasons for the score perhaps with examples of the difficulties this creates and what should be done.
Admin Budget (FA_Budg): The stakeholder's assessment of the adequacy of the TSP's budgetary provision for its own finance and administration function.
Date and time stamp. When this property was last amended.
Score out of 100. The stakeholder's latest assessment of this property.
Explanation. The stakeholder's reasons for the score perhaps with examples of the difficulties this creates and what should be done.
Admin Leadership (FA_Lead): The stakeholder's assessment of the adequacy of the TSP's leadership arrangements for managing its finance and administration.
Date and time stamp. When this property was last amended.
Score out of 100. The stakeholder's latest assessment of this property.
Explanation. The stakeholder's reasons for the score perhaps with examples of the difficulties this creates and what should be done.
DIMENSION: People and Organisation (HR&ORG)
How the stakeholder perceives the effectiveness of the people, how they are organized, the culture of the organisation, and the HR management aspects of the technology service provider and the impact of this on the business.
Date and time stamp. When People and Organisation was last amended, i.e. not the date it was last incorporated unchanged into an overall assessment of the TBE.
People and Organisation score out of 100. The rating is calculated as the arithmetic mean of the latest scores given by the stakeholder for each of the properties below, at the date and time stated above, including those carried forward unchanged from a previous assessment.
Long description. Any comment which helps to explain the assessment calculated for this dimension or point the reader to the property (below) that is primarily responsible where a fuller explanation is available for the change in the score above.
PROPERTIES of People and Organisation:
Organisational Structure and Accountabilities (HO_Org): The stakeholder's assessment of the manner in which the TSP organises its people and defines their accountabilities, the apparent effectiveness of this, and how this impacts the business.
Date and time stamp. When this property was last amended.
Score out of 100. The stakeholder's latest assessment of this property.

Explanation. The stakeholder's reasons for the score perhaps with examples of the difficulties this creates and what should be done.
Culture and Motivation (HO_Cult): The stakeholder's assessment of the culture and motivation across technology services and its supply chain, i.e. the belief systems or values that are driving behaviour and attitudes and how this affects the business, e.g. collaboration, decision-making, direction, and delegation, etc.
Date and time stamp. When this property was last amended.
Score out of 100. The stakeholder's latest assessment of this property.
Explanation. The stakeholder's reasons for the score perhaps with examples of the difficulties this creates and what should be done.
Training and Development (HO_Tr&Dev): The stakeholder's assessment of the training and development given to technology services people as a response to skills gaps, succession planning, and personal development.
Date and time stamp. When this property was last amended.
Score out of 100. The stakeholder's latest assessment of this property.
Explanation. The stakeholder's reasons for the score perhaps with examples of the difficulties this creates and what should be done.
Individual and Group Effectiveness (HO_I&Geff): The stakeholder's assessment of the apparent effectiveness of the process and reward systems for TSP people for assessing and improving the performance of the individual and the group.
Date and time stamp. When this property was last amended.
Score out of 100. The stakeholder's latest assessment of this property.
Explanation. The stakeholder's reasons for the score perhaps with examples of the difficulties this creates and what should be done.
Recruitment (HO_Recr): The stakeholder's assessment of the suitability of the TSP's process for recruitment and its effectiveness for meeting the needs of the business.
Date and time stamp. When this property was last amended.
Score out of 100. The stakeholder's latest assessment of this property.
Explanation. The stakeholder's reasons for the score perhaps with examples of the difficulties this creates and what should be done.
People and Organisation Budget (HO_Budg): The stakeholder's assessment of the adequacy of the TSP's budget for human resource management.
Date and time stamp. When this property was last amended.
Score out of 100. The stakeholder's latest assessment of this property.
Explanation. The stakeholder's reasons for the score perhaps with examples of the difficulties this creates and what should be done.
People and Organisation Leadership (HO_Lead): The stakeholder's assessment of the adequacy of the TSP's leadership of human resource management.
Date and time stamp. When this property was last amended.
Score out of 100. The stakeholder's latest assessment of this property.
Explanation. The stakeholder's reasons for the score perhaps with examples of the difficulties this creates and what should be done.

DIMENSION: Design and Definition of Products and Services (DESDEF)
How the stakeholder perceives the effectiveness of the activities of the TSP that involve establishing the technology requirements, the design, the technical specification of the solution, and the impact of this on the business.
Date and time stamp. When Design and Definition of Products and Services was last amended, i.e. not the date it was last incorporated unchanged into an overall assessment of the TBE.
Design and Definition of Products and Services score out of 100. The rating is calculated as the arithmetic mean of the latest scores given by the stakeholder for each of the properties below, at the date and time stated above, including those carried forward unchanged from a previous assessment.
Long description. Any comment which helps to explain the assessment calculated for this dimension or point the reader to the property (below) that is primarily responsible where a fuller explanation is available for the change in the score above.
PROPERTIES of Design and Definition of Products and Services:
Understanding the Requirement (DD_Req). The stakeholder's assessment of the accuracy of the TSP's interpretation of the business requirement explicit in the technology solutions being used by the stakeholder.
Date and time stamp. When this property was last amended.
Score out of 100. The stakeholder's latest assessment of this property.
Explanation. The stakeholder's reasons for the score perhaps with examples of the difficulties this creates and what should be done.
Product and Service Design (DD_Des). The stakeholder's assessment of the TSP's design capabilities as they relate to its products and services, including the stakeholder interface.
Date and time stamp. When this property was last amended.
Score out of 100. The stakeholder's latest assessment of this property.
Explanation. The stakeholder's reasons for the score perhaps with examples of the difficulties this creates and what should be done.
Build Versus Buy Decisions (DD_BvsB). The stakeholder's assessment of how well the TSP is able to decide between investing in industry standard vendor solutions and bespoke development with or without partner organisations.
Date and time stamp. When this property was last amended.
Score out of 100. The stakeholder's latest assessment of this property.
Explanation. The stakeholder's reasons for the score perhaps with examples of the difficulties this creates and what should be done.
Design and Definition Budget (DD_Budg). The stakeholder's assessment of the adequacy of the TSP's budget for managing the design and definition activities.
Date and time stamp. When this property was last amended.
Score out of 100. The stakeholder's latest assessment of this property.
Explanation. The stakeholder's reasons for the score perhaps with examples of the difficulties this creates and what should be done.

Design and Definition Leadership (DD_Lead). The stakeholder's assessment of the adequacy of the TSP's leadership arrangements for managing the design and definition activities.
Date and time stamp. When this property was last amended.
Score out of 100. The stakeholder's latest assessment of this property.
Explanation. The stakeholder's reasons for the score perhaps with examples of the difficulties this creates and what should be done.
DIMENSION: Development and Operational Delivery of Products and Services (DEVOPS):
How the stakeholder perceives the effectiveness of the activities of the TSP involving the acquisition or development of solutions and services, their deployment, operation and support services, and the impact of this on the business.
Date and time stamp. When Development and Operational Delivery of Products and Services was last amended, i.e. not the date it was last incorporated unchanged into an overall assessment of the TBE.
Development and Operational Delivery of Products and Services score out of 100. The rating is calculated as the arithmetic mean of the latest scores given by the stakeholder for each of the properties below, at the date and time stated above, including those carried forward unchanged from a previous assessment.
Long description. Any comment which helps to explain the assessment calculated for this dimension or point the reader to the property (below) that is primarily responsible where a fuller explanation is available for the change in the score above.
PROPERTIES of Development and Operational Delivery of Products and Services:
Program and Project Management (DO_PM). The stakeholder's assessment of the TSP's ability to manage programs or projects to deliver products and services (development and deployment) to plan, to a prescribed outcome, i.e. to a requirement and design, working as expected, and achieving the approval of the sponsoring stakeholder group. This includes: scope, cost, timescales, deliverables, quality outcomes, reporting and communications, change control, etc.
Date and time stamp. When this property was last amended.
Score out of 100. The stakeholder's latest assessment of this property.
Explanation. The stakeholder's reasons for the score perhaps with examples of the difficulties this creates and what should be done.
Data Quality (DO_DataQ). The stakeholder's assessment of the quality of the data input to and available from systems and services.
Date and time stamp. When this property was last amended.
Score out of 100. The stakeholder's latest assessment of this property.
Explanation. The stakeholder's reasons for the score perhaps with examples of the difficulties this creates and what should be done.
Software Engineering (DO_SE). The stakeholder's assessment of the TSP's proficiency in the software engineering aspects of development and operational delivery activity, i.e. coding, testing, systems integration, and the customisation/configuration of bought-in solutions and services, etc.

Date and time stamp. When this property was last amended.
Score out of 100. The stakeholder's latest assessment of this property.
Explanation. The stakeholder's reasons for the score perhaps with examples of the difficulties this creates and what should be done.
Deployment (DO_Depl). The stakeholder's assessment of the TSP's proficiency in the deployment of newly tested solutions into the business.
Date and time stamp. When this property was last amended.
Score out of 100. The stakeholder's latest assessment of this property.
Explanation. The stakeholder's reasons for the score perhaps with examples of the difficulties this creates and what should be done.
Service Operations Management (DO_SvOps). The stakeholder's assessment of the TSP's proficiency in the day-to-day delivery of operational services to the business against key measures such as reliability, availability, meantime to failure, mean time to repair, unplanned unavailability, loss of data, reliability of connections, etc.
Date and time stamp. When this property was last amended.
Score out of 100. The stakeholder's latest assessment of this property.
Explanation. The stakeholder's reasons for the score perhaps with examples of the difficulties this creates and what should be done.
Upgrades and Change Control (DO_CC). The stakeholder's assessment of the TSP's routine change-control procedures for its products and services for version control, and to maintain the currency of release levels and contractual arrangements.
Date and time stamp. When this property was last amended.
Score out of 100. The stakeholder's latest assessment of this property.
Explanation. The stakeholder's reasons for the score perhaps with examples of the difficulties this creates and what should be done.
Support Services (DO_SS). The stakeholder's assessment of the operation of technical support services in the form of help/service desks including call frequency, closure rates, problem resolution, timeliness of communication, etc.
Date and time stamp. When this property was last amended.
Score out of 100. The stakeholder's latest assessment of this property.
Explanation. The stakeholder's reasons for the score perhaps with examples of the difficulties this creates and what should be done.
Quality Management (DO_QM). The stakeholder's assessment of the effectiveness of the quality assurance processes and outcomes to ensure fitness for purpose from design to delivery.
Date and time stamp. When this property was last amended.
Score out of 100. The stakeholder's latest assessment of this property.
Explanation. The stakeholder's reasons for the score perhaps with examples of the difficulties this creates and what should be done.
Delivery Budget (DO_Budg). The stakeholder's assessment of the adequacy of the TSP's budgetary provision for managing the development and operational delivery activities.
Date and time stamp. When this property was last amended.

Score out of 100. The stakeholder's latest assessment of this property.
Explanation. The stakeholder's reasons for the score perhaps with examples of the difficulties this creates and what should be done.
Delivery Leadership (DO_Lead). The stakeholder's assessment of the adequacy of the TSP's leadership arrangements for managing the development and operational delivery activities.
Date and time stamp. When this property was last amended.
Score out of 100. The stakeholder's latest assessment of this property.
Explanation. The stakeholder's reasons for the score perhaps with examples of the difficulties this creates and what should be done.
DIMENSION: Security (SECUR):
How the stakeholder perceives the effectiveness of the TSP in security provision when compared with industry best practice, legislation, or its effectiveness in protecting the assets of the business and others from attack and exploitation. The score is calculated as a mean of the scores for the properties below.
Date and time stamp. When Security was last amended, i.e. not the date it was last incorporated unchanged into an overall assessment of the TBE.
Security score out of 100. The rating is calculated as the arithmetic mean of the latest scores given by the stakeholder for each of the properties below, at the date and time stated above, including those carried forward unchanged from a previous assessment.
Long description. Any comment which helps to explain the assessment calculated for this dimension or point the reader to the property (below) that is primarily responsible where a fuller explanation is available for the change in the score above.
PROPERTIES of Security:
Site and Business Security (Sec_Site). The stakeholder's assessment of the TSP's safeguarding arrangements for protecting the business itself and its operational and administrative sites together with any supporting processes and systems. Key aspects include security policy, physical and environmental site security, asset management, selection of personnel and training in social engineering, identity security, visitor security, emergency management.
Date and time stamp. When this property was last amended.
Score out of 100. The stakeholder's latest assessment of this property.
Explanation. The stakeholder's reasons for the score perhaps with examples of the difficulties this creates and what should be done.
Information/Access Security (Sec_Info). The stakeholder's assessment of the effectiveness of the TSP's largely digital safeguards taken to protect unauthorised access to information, systems, and networks. Key aspects include: processor security, device security, network and web security, maintenance patches and updates, loss of data including accidental disclosure through social engineering/phishing activity, data access controls, identification safeguards, information systems (acquisition, development and maintenance), incident management, any additional compliance and regulatory requirements.

Date and time stamp. When this property was last amended.
Score out of 100. The stakeholder's latest assessment of this property.
Explanation. The stakeholder's reasons for the score perhaps with examples of the difficulties this creates and what should be done.
Business Continuity/Disaster Recovery (Sec_BC/DR). The stakeholder's assessment of the ability of the TSP to recover the technology services in a timely and appropriate manner following a failure and the arrangements for continuing to operate the business during the interim period and thereafter.
Date and time stamp. When this property was last amended.
Score out of 100. The stakeholder's latest assessment of this property.
Explanation. The stakeholder's reasons for the score perhaps with examples of the difficulties this creates and what should be done.
Security Budget (Sec_Budg). The stakeholder's assessment of the adequacy of the TSP's budget for managing the security aspects of the technology services.
Date and time stamp. When this property was last amended.
Score out of 100. The stakeholder's latest assessment of this property.
Explanation. The stakeholder's reasons for the score perhaps with examples of the difficulties this creates and what should be done.
Security Leadership (Sec_Lead). The stakeholder's assessment of the adequacy of the TSP's leadership arrangements for managing the security aspects of the technology services.
Date and time stamp. When this property was last amended.
Score out of 100. The stakeholder's latest assessment of this property.
Explanation. The stakeholder's reasons for the score perhaps with examples of the difficulties this creates and what should be done.
DIMENSION: Technical Infrastructure (TECHINF)
How the stakeholder perceives the effectiveness of the TSP in maintaining the utility value of the technical infrastructure (hardware, software, and telecommunications) of the business and the impact of its activities on the environment.
Date and time stamp. When Technical Infrastructure was last amended, i.e. not the date it was last incorporated unchanged into an overall assessment of the TBE.
Technical Infrastructure score out of 100. The rating is calculated as the arithmetic mean of the latest scores given by the stakeholder for each of the properties below, at the date and time stated above, including those carried forward unchanged from a previous assessment.
Long description. Any comment which helps to explain the assessment calculated for this dimension or point the reader to the property (below) that is primarily responsible where a fuller explanation is available for the change in the score above.
PROPERTIES of Technical Infrastructure:
Scope of the Vision (TI_Vision). The stakeholder's assessment of the TSP's vision for the infrastructure that supports the emerging and future needs of the extended business enterprise. Key considerations include: compatibility, i.e. the homogeneity across internal and external value networks, reach, connectivity,

availability, speed, response times, reliability, and the richness (diversity and depth) of high value features.
Date and time stamp. When this property was last amended.
Score out of 100. The stakeholder's latest assessment of this property.
Explanation. The stakeholder's reasons for the score perhaps with examples of the difficulties this creates and what should be done.
Skills Training for the Business (TI_SkTr). The stakeholder's assessment of the training available to stakeholders that helps the business to get the best out of the technology already deployed.
Date and time stamp. When this property was last amended.
Score out of 100. The stakeholder's latest assessment of this property.
Explanation. The stakeholder's reasons for the score perhaps with examples of the difficulties this creates and what should be done.
Introducing New Technology (TI_Intro). The stakeholder's assessment of the philosophy and means by which new technology is selected by the TSP and how well that fits with the culture of the business.
Date and time stamp. When this property was last amended.
Score out of 100. The stakeholder's latest assessment of this property.
Explanation. The stakeholder's reasons for the score perhaps with examples of the difficulties this creates and what should be done.
Asset Register/Technology Deployed (TI_AReg). The stakeholder's assessment of the accuracy and completeness of the TSP's maintenance of the register of technology assets deployed, including their configuration, support status, location, warranties and licenses, etc.
Date and time stamp. When this property was last amended.
Score out of 100. The stakeholder's latest assessment of this property.
Explanation. The stakeholder's reasons for the score perhaps with examples of the difficulties this creates and what should be done.
Ecological and Environmental Footprint (TI_EEF). The stakeholder's assessment of the impact that the TSP has on the ecological or environmental footprint of the business, for example natural resources used or harmful gases/other waste produced.
Date and time stamp. When this property was last amended.
Score out of 100. The stakeholder's latest assessment of this property.
Explanation. The stakeholder's reasons for the score perhaps with examples of the difficulties this creates and what should be done.
Technical Infrastructure Budget (TI_Budg). Score out of 100, the adequacy of the TSP's budgetary provision for managing the technical infrastructure supporting the services.
Date and time stamp. When this property was last amended.
Score out of 100. The stakeholder's latest assessment of this property.
Explanation. The stakeholder's reasons for the score perhaps with examples of the difficulties this creates and what should be done.

Technical Infrastructure Leadership (TI_Lead). The stakeholder's assessment of the adequacy of the TSP's leadership arrangements for managing the technical infrastructure associated with the services.
Date and time stamp. When this property was last amended.
Score out of 100. The stakeholder's latest assessment of this property.
Explanation. The stakeholder's reasons for the score perhaps with examples of the difficulties this creates and what should be done.
DIMENSION: Bought-In (Supplied) Technology Products and Services (BITP&S)
How the stakeholder perceives the effectiveness of third-party suppliers of technology products or services with whom the main TSP, or part of the business, has a supply contract, e.g. for hardware, software, or services.
Date and time stamp. When Bought-In Technology Products and Services was last amended, i.e. not the date it was last incorporated unchanged into an overall assessment of the TBE.
KSG of Third-party Supplier. Identifies the specific third-party supplier from the KSG graphic, if shown and if known.
Bought-In Technology Products and Services score out of 100. The rating is calculated as the arithmetic mean of the latest scores given by the stakeholder for each of the properties below, at the date and time stated above, including those carried forward unchanged from a previous assessment.
Long description. Any comment which helps to explain the assessment calculated for this dimension or point the reader to the property (below) that is primarily responsible where a fuller explanation is available for the change in the score above.
PROPERTIES of Bought-In (Supplied) Technology Products and Services:
Supplier's Business Awareness (B_Baware). The stakeholder's assessment of the adequacy of Business Awareness as it applies to third-party technology service providers.
Date and time stamp. When this property was last amended.
Score out of 100. The stakeholder's latest assessment of this property.
Explanation. The stakeholder's reasons for the score perhaps with examples of the difficulties this creates and what should be done. The properties from the Business Awareness dimension may be used as a prompt when commenting.
Supplier's Marketing/Communications (B_Marcom). The stakeholder's assessment of the adequacy of Marketing and Communications as it applies to third-party technology service providers.
Date and time stamp. When this property was last amended.
Score out of 100. The stakeholder's latest assessment of this property.
Explanation. The stakeholder's reasons for the score perhaps with examples of the difficulties this creates and what should be done. The properties from the Marketing/Communications dimension may be used as a prompt when commenting.
Supplier's Sales/Value Add (B_ValAdd). The stakeholder's assessment of the adequacy of Sales/Value Add as it applies to third-party technology service providers.

Date and time stamp. When this property was last amended.
Score out of 100. The stakeholder's latest assessment of this property.
Explanation. The stakeholder's reasons for the score perhaps with examples of the difficulties this creates and what should be done. The properties from the Sales/ Value Add dimension may be used as a prompt when commenting.
Supplier's Commercial Administration (B_F&Admin). The stakeholder's assessment of the adequacy of commercial finance and administration as it applies to third-party technology service providers.
Date and time stamp. When this property was last amended.
Score out of 100. The stakeholder's latest assessment of this property.
Explanation. The stakeholder's reasons for the score perhaps with examples of the difficulties this creates and what should be done. The properties from the Finance and Administration dimension may be used as a prompt when commenting.
Supplier's People and Organisation (B_HR&Org). The stakeholder's assessment of the adequacy of People and Organisation as it applies to third-party technology service providers.
Date and time stamp. When this property was last amended.
Score out of 100. The stakeholder's latest assessment of this property.
Explanation. The stakeholder's reasons for the score perhaps with examples of the difficulties this creates and what should be done. The properties from the People and Organisation dimension may be used as a prompt when commenting.
Supplier's Design and Definition of Products and Services (B_DesDef). The stakeholder's assessment of the adequacy of Design and Definition of Products And Services as it applies to third-party technology service providers.
Date and time stamp. When this property was last amended.
Score out of 100. The stakeholder's latest assessment of this property.
Explanation. The stakeholder's reasons for the score perhaps with examples of the difficulties this creates and what should be done. The properties from the Design and Definition of Products and Services dimension may be used as a prompt when commenting.
Supplier's Development and Operational Delivery of Products and Services (B_DevOps). The stakeholder's assessment of the adequacy of Development and Operational Delivery of Products and Services as it applies to third-party technology service providers.
Date and time stamp. When this property was last amended.
Score out of 100. The stakeholder's latest assessment of this property.
Explanation. The stakeholder's reasons for the score perhaps with examples of the difficulties this creates and what should be done. The properties from the Development and Operational Delivery of Products and Services dimension may be used as a prompt when commenting.
Supplier's Security (B_Secur). The stakeholder's assessment of the adequacy of Security as it applies to third-party technology service providers.
Date and time stamp. When this property was last amended.
Score out of 100. The stakeholder's latest assessment of this property.

Explanation. The stakeholder's reasons for the score perhaps with examples of the difficulties this creates and what should be done. The properties from the Security dimension may be used as a prompt when commenting.
Supplier's Technical Infrastructure (B_TechInf). The stakeholder's assessment of the adequacy of the Technical Infrastructure of third-party technology service providers and their impact on the environment.
Date and time stamp. When this property was last amended.
Score out of 100. The stakeholder's latest assessment of this property.
Explanation. The stakeholder's reasons for the score perhaps with examples of the difficulties this creates and what should be done. The properties from the Technical Infrastructure dimension may be used as a prompt when commenting.
Supplier's Supplied (Bought-In) Technology Products and Services (B_Bitp&s). The stakeholder's assessment of the adequacy of products and services bought-in by third-party technology service providers which are subsequently supplied to the business.
Date and time stamp. When this property was last amended.
Score out of 100. The stakeholder's latest assessment of this property.
Explanation. The stakeholder's reasons for the score perhaps with examples of the difficulties this creates and what should be done. The same properties as used in this Bought-In (Supplied) Technology Products and Services dimension may be used as a prompt when commenting.
Supplier's Budget (B_Budg). The stakeholder's assessment of the adequacy of the TSP's budgetary provision for managing third-party technology service providers (the procurement budget).
Date and time stamp. When this property was last amended.
Score out of 100. The stakeholder's latest assessment of this property.
Explanation. The stakeholder's reasons for the score perhaps with examples of the difficulties this creates and what should be done.
Supplier's Leadership (B_Lead). The stakeholder's assessment of the adequacy of the TSP's leadership arrangements for the procurement and management of third-party technology service providers.
Date and time stamp. When this property was last amended.
Score out of 100. The stakeholder's latest assessment of this property.
Explanation. The stakeholder's reasons for the score perhaps with examples of the difficulties this creates and what should be done.

Table A.5 Data fields for Service Requirement

CATEGORY: Service Requirement
A stakeholder's suggestion as to what next steps need to be taken to resolve the issues identified.
Header data. See Table A.1.
Date and time stamp. When the entry for the Service Requirement was made/amended.

Links. Connects to any TBE assessment, Business Need, and/or Change that is considered to be a Causal Link.
Short description. Name or short phrase used to describe the work that the stakeholder suggests needs to be undertaken. More information is entered using one or more of the service types in the Dimensions listed below to describe what actions are needed to address the issues identified.
Status. Valid (blank), Invalid (X), or In Need of Checking (?). Note that Status is a feature of 'Hindsight' and nothing is actually deleted.
Reason for status Change. Stakeholder's reason for changing the status.
DIMENSION: Overall Service Management
Date and time stamp. When the entry was made.
Long Description. The stakeholder's description of any requirement relating to the overall management of services.
Tentative Man Day Estimate: Entered by the Gold Team
Date and time stamp. When the entry was made by the Gold Team.
Outcome: Follow up and recommendations.
Date and Time Stamp: When the follow-up entry was made by the Gold Team.
DIMENSION: Business Planning
Date and time stamp. When the entry was made.
Long Description. The stakeholder's description of the business planning services thought to be required. This may include assistance with business analysis, business/technology strategy, business engineering/re-engineering, and business process engineering, etc., i.e. any business transformation requirement arising from the need for a new business model or to modify/introduce new value networks.
Tentative Man Day Estimate: Entered by the Gold Team
Date and time stamp. When the entry was made by the Gold Team.
Outcome: Follow up and recommendations.
Date and Time Stamp: When the follow-up entry was made by the Gold Team.
DIMENSION: Application/Digital Services
Date and time stamp. When the entry was made.
Long Description. The stakeholder's description of the application or digital services that are thought to be required. This may include any design, definition, and software engineering or the operation, discovery, and use of technology products and services.
Tentative Man Day Estimate: Entered by the Gold Team
Date and time stamp. When the entry was made by the Gold Team.
Outcome: Follow up and recommendations.
Date and Time Stamp: When the follow-up entry was made by the Gold Team.
DIMENSION: Transition Services
Date and time stamp. When the entry was made.
Long Description. The stakeholder's description of the transition services that are thought to be required including deployment and assistance with actual business process change and product/service implementation.

Tentative Man Day Estimate: Entered by the Gold Team
Date and time stamp. When the entry was made by the Gold Team.
Outcome: Follow up and recommendations.
Date and Time Stamp: When the follow-up entry was made by the Gold Team.
DIMENSION: Infrastructure Services
Date and time stamp. When the entry was made.
Long Description. The stakeholder's description of the infrastructure services that are thought to be required including the central and distributed hardware, software, communications, and other services necessary to allow the technology infrastructure to support the stated needs.
Tentative Man Day Estimate: Entered by the Gold Team
Date and time stamp. When the entry was made by the Gold Team.
Outcome: Follow up and recommendations.
Date and Time Stamp: When the follow-up entry was made by the Gold Team.

Table A.6 Data fields for Alignment and Governance

CATEGORY: Alignment and Governance:
Actions to be taken as a result of stakeholder requests and Gold Team analysis and investigation.
Header data. See Table A.1—but in this case the stakeholder must be a member of the Gold Team at the time the entry is made.
Date and time stamp. When the entry was made.
Links. Connects to all Changes, Business Needs, TBE assessments, or Service Requirements that are considered to be a Causal Link.
Short description. Name or short phrase provided by the Gold Team to be used to identify this alignment intervention. More information is provided in the long description of each related alignment task below.
Status. Valid (blank), Invalid (X), or In Need of Checking (?). See Hindsight.
Reason for status Change. Stakeholder's reason for changing the status.
ALIGNMENT TASK:
The planned response(s) to the above.
Date and time stamp. When the entry was made.
Long description. The Gold Team's description of a response or intervention designed to address a Business Need or bring the assessment of the TBE closer to what the business stakeholders consider to be ideal. A cross-reference is permitted to any external information source that will provide access to more detailed data, e.g. detailed proposals and project plans submitted by the TSP.
WORK SCHEDULE: The progress of each task in the work schedule is shown below.
Date when need first identified: date of and link to the first entry that identified the need for the eventual Alignment Task
Date of preliminary study: the date on which resource was assigned to look at the problems being raised in order to formulate the response.
Date work proposed: the date on which the response to the problems was proposed.

Date work approved: the date on which the proposed response was approved.
Date work to be started: the planned date when work is due to start on the proposed and approved response.
Date ready: the date on which the completed work is expected to be ready for deployment.
Date deployed/completed: the date on which the completed work was fully deployed or implemented.
GOVERNANCE:
The two key measures of the effectiveness of the outcomes of Alignment Tasks are Agility and Closeness, as defined below. Entries for these measures of governance will be entered by the Gold Team and be independently verifiable by scrolling through historical data using the features of the TBE Radar Diagram.
Date and time stamp. When the entry was made.
KSG. The KSG for which the metrics (below) apply. Note that a single alignment task may address a spectrum of requirements across a number of KSGs and address each to different degree. The Gold Team may choose to select the KSG principally affected or include a selection of KSGs where the problem and solution are widely applicable.
AGILITY: Best expression of the perceived speed of response for a KSG by the service provider in provisioning the Alignment Task.
Agility elapsed days. Speed of response (i.e. the elapsed time using the dates below).
Date when need first identified within the KSG.
Date of worst assessment for KSG due to the problem reported.
Date alignment task deployed/completed.
CLOSENESS: The improvement in overall alignment being reported by stakeholders (by comparing the pre- and post-deployment assessments of the TBE at the dates used in Agility, above.) for the KSGs that were affected by the Change, etc. The worst and best scores and entry dates will be recorded here for every Alignment Task for easy reference so that improvement can be continuously monitored.
Worst rating for KSG due to problem reported.
Best rating for KSG post deployment/completion.
Notes. Any notes of relevance including mitigating circumstance, patterns across KSGs, contagion, mean scores compared with worst and best, explanations for the improvement, or lack of, stakeholder comments pre- and post-implementation, etc.

Glossary

Active Governance The continuous application of tests for Closeness of Alignment and Agility of Response and other tests of alignment.

Advanced manufacturing and processing To create new kinds of products that can't be made cost effectively with conventional processes.

Advanced materials Materials with engineered properties.

Agile Manifesto A document that sets out the key values and principles behind the Agile philosophy.

Agility of Response A governance term that represents the time between a Change being identified or notified and an Alignment Task being successfully deployed which addresses the Spectrum of Business Requirement that is consequential to the Change.

AI Artificial Intelligence is intelligence programmed into machines that mimics natural intelligence

Autonomic computing Self-managing characteristics of distributed computing resources.

BAWARE Abbreviation for Business Awareness, e.g. as used in the TBE radar diagram

Big Data Very large data sets that can reveal patterns, trends, and associations.

BITP&S Abbreviation for Bought-in Technology Products and Services, e.g. as used in the TBE radar diagram

Blockchain A mechanism for maintaining records of transactions in cryptocurrencies and in other distributed applications where the integrity of the data must be maintained.

BTRM Business and Technology Relationship Model. Categories are Change, Business Need, Total Business Experience, Service Requirement, and Alignment and Governance.

Business technology Any technology that supports a business. Used here when talking about the technologies that will have to be managed going forward, including current technologies and legacy systems.

Causal Chain A complete Causal Chain consists of the categories Change, Business Need, Total Business Experience, and Service Requirement, and their linkages.

D. Miller, *Managing Agile Business Technology*, Management for Professionals,
https://doi.org/10.1007/978-3-030-90598-9

CEO Chief Executive Officer

CIO Chief Information Officer

Closeness of Alignment A governance term based on the state of the Spectrum and Business Requirement and it represents the difference between what exists and what is needed, or the degree of alignment.

CMDB Configuration Management Database. A database used to store information about hardware and software assets broken down into logical layers, i.e. similar to assembly and component breakdowns.

CMMI and ISO 15504 Stands for Capability Maturity Model Integration. Supported by ISO/IEC 15504 and replacements.

CMO Chief Marketing Officer

Consumers External stakeholders who are end customers of business products or services who are direct or indirect users of the organisation's technology products or services.

Contagion Contagion is using past data to help us to anticipate the future—it can provide foresight. It shows how the impact of a Change spreads throughout an organisation from the early signs that a Change is happening to experiencing the full effect.

C-Suite The executive team with Chief titles, e.g. Chief Executive Officer, Chief Financial Officer, Chief Information Officer, etc.

CTO Chief Technology Officer

Current IT delivery model The IT delivery model as exemplified by current practices. The term, IT delivery model, was used throughout the research to refer to the processes used to design and build systems until during the later development phases it became clear that the outcomes from the research were more widely applicable to the management and governance of all business technology.

Customers/Business customers External stakeholders who are employed by trade customers of a business product or service who are direct or indirect users of the organisation's technology products or services. Customer is also used generically for a trade buyer of product or service in any context, e.g. when discussing sales/marketing activity.

DESDEF Abbreviation for Design &Definition of Products and Services, e.g. as used in the TBE radar diagram

DEVOPS Abbreviation for Development and Operational Delivery of Products and Services, e.g. as used in the TBE radar diagram

Disconfirmation Model A measure of service quality (satisfaction) based on performance minus expectation but the model is criticised for not providing adequate information for decision makers as to the causes of dissatisfaction.

EA Enterprise Architecture. A blueprint for design that captures the key data and process components of an enterprise as captured or conceived by the business analyst. It may be conceptual and connect to the physical configuration management database CMDB.

F&ADMIN Abbreviation for Finance and Administration, e.g. as used in the TBE radar diagram

Genchi Genbutsu From the Toyota 'go and see' list, i.e. business process that needs to be looked at in the light of what is being entered into the Active Management and Governance system.

Gold Team The part of the Executive, or a team accountable to the Executive, who monitors the Closeness of Alignment, instigates Alignment Tasks, and monitors the Agility of Response.

GTM Grounded Theory Method, a qualitative research method for systematically discovering or constructing theory from data using comparative analysis.

Hindsight A process for maintaining the integrity of historic data, for looking back at a past situation, e.g. to one where the Change or Business Need was assumed to be correct, in order to understand with hindsight our past actions, whilst simultaneously entering more recent and more accurate information for actively managing the situation going forward.

HR&ORG Abbreviation for People, Organisation and Culture, e.g. as used in the TBE radar diagram

ICT Information and communications technology

Internet of Things (IoT) The network of devices embedded with technology that can connect and exchange data with other devices and systems over the Internet.

ISO International Standards Organisation

IT Information Technology. Used here when referring to traditional computer technologies used in the past and to the present day to process information

IT governance committee A committee appointed by the board of directors to whom the task of IT governance was delegated

JIT decision-making Investment ahead of time or on time facilitated via Alignment and Governance which is a dynamic key control point for the business when using Active Management and Governance.

KSG Key Stakeholder Group. An organisational grouping associated with a discreet business mission or group of activities/processes and hence having a unique role in the extended enterprise. A chart of the extended enterprise can also include KSGs for TSPs, key suppliers, and key customers. Does not replace the internal organisation chart which often has historical relevance and is based around people, reporting lines, seniority, or can be more transitory.

Likert scale A type of psychometric response scale typically using five levels, e.g. (1) Strongly disagree; (2) Disagree; (3) Neither agree nor disagree; (4) Agree; (5) Strongly agree.

MARCOM Abbreviation for Marketing and Communications, e.g. as used in the TBE radar diagram

Modus operandi Used here to describe the cocktail of current frameworks, methodologies, and standards selected as management tools in any given situation.

Moore's Law Moore's law is an observation and projection of an historical trend that states that the number of transistors on a microchip (and hence computing power) doubles about every two years and halves in price.

MVP Minimum viable product is a term used when creating something that will prove the concept but with the least effort.

Multi-agent systems/MAS A multi-agent system consists of multiple decision-making agents which interact in a shared environment to achieve common or conflicting goals.

Nanotechnologies The manipulation of materials on an atomic or molecular scale to build microscopic devices or to control the structure of matter.

Organisation/Business Any organisation in the context in which it is used but may be referring here to the client enterprise that is managing business technology or making governance judgements.

PDCA An iterative planning/management cycle standing for Plan, Do, Check, Act. It has many variations and is sometimes referred to as the Shewhart Cycle or the Deming Cycle. It is the basis for many of the original IT methodologies.

Quantum Computing Performs computation by exploiting different states to conventional computers which results in faster computational times.

RACI A responsibility assignment chart standing for responsible, accountable, consulted, and informed. Describes how people participate in a task/project.

SECUR Abbreviation for Security, e.g. as used in the TBE radar diagram

Self-adaptive and self-managing systems Computer systems that will sense a change and adapt to it automatically.

Service catalogue A database of IT service offerings that form the TSP's service portfolio.

Service dominant logic Focus on service, co-creation of value, and value-in-exchange, rather than the units of output that are traded. The logic can be extended to services operating in an end-to-end manner across communities, value constellations or value networks rather than value chains, and can be conceived as systematic social innovation involving the continuous design and redesign of complex business systems, as facilitated by Active Management and Governance.

Six Sigma A management technique to improve business processes with the aim of reducing the probability that an error or defect will occur. Mathematically it would take a six-standard-deviation event from the mean for an error to occur.

SLA Service Level Agreement. Defines the services expected by a customer from a supplier and what performance metrics will be used.

SOA Service Oriented Architecture. A way to make software components reusable via service interfaces so they can be rapidly incorporated into new applications.

Spectrum of Business Requirement Business requirements can be expressed in different ways and at different times by different people. Together Business Need, the Total Business Experience, and the Service Requirement can be referred to as the 'Spectrum of Business Requirement',

Sprint A sprint in agile methods terms is a short time limited period within which a set piece of work has to be completed.

Stakeholder Any user, customer, consumer, supplier, business person, or TSP member defined within the system and belonging to a relevant Key Stakeholder Group (KSG).

Super group A KSG group equivalent to a divisional grouping that is used to summarise the TBE assessments by type of external entity, e.g. all suppliers, trade customers, etc.

Suppliers External stakeholders who are employed by a supplier to the business of products and services, other than business technology, and who are direct or indirect users of the organisation's business technology.

TBE Total Business Experience (see BTRM).

TECHINF Abbreviation for Technical Infrastructure, e.g. as used in the TBE radar diagram

Third-party Supplier A Third-party Supplier of technology products or services (or Third-party TSP) with whom a TSP, or part of the business, has a supply contract.

TOGAF A content framework for enterprise architectures.

TQM Total Quality Management. A quality management approach to process improvement.

TSP Technology Service Provider. Will be contracted directly to the enterprise, or part of it. The TSP may be the internal IT organisation, an outsourced former internal IT organisation, an external TSP, or one of a number of TSPs. Also, see 'Third-part Supplier'.

TSP member A stakeholder who is employed by a TSP.

UML Unified Modelling Language used to visualise a system specification or structure.

Users Stakeholders who are users of the technology products or services being managed or assessed.

VALADD Abbreviation for Sales and Value Add, e.g. as used in the TBE radar diagram

Value-in-use Is tested continuously via TBE Assessments as a means of co-creating value and deriving a proxy for value-in-exchange

Index

D. Miller, *Managing Agile Business Technology*, Management for Professionals,
https://doi.org/10.1007/978-3-030-90598-9

Préface

Natif de Bretagne, non loin de la baie de Saint-Brieuc, je voue une passion gourmande à la noix de Saint-Jacques, ce bonbon de la mer, qui se prête facilement aux facéties de tous les cuisiniers, qu'ils soient amateurs ou professionnels.

Je connais peu de personnes qui n'apprécient pas les noix de Saint-Jacques, avec ou sans leur corail – que je n'utilise que très rarement, sauf mixé pour obtenir une crème onctueuse.

En revanche, j'en connais beaucoup trop qui la consomment en dehors des campagnes de pêche.
La Saint-Jacques, qu'elle provienne de la baie de Saint-Brieuc, de Normandie où elle affiche avec fierté un Label Rouge ou encore de la côte atlantique, ne se pêche que 7 mois de l'année, d'octobre à avril. Rendons grâce aux pêcheurs qui ont accepté la limitation des volumes et du temps de pêche pour nous permettre de l'attendre chaque année avec cette impatience propre aux gourmets préférant de loin les produits frais et de saison.

Froide, chaude, entière ou émincée, la noix de Saint-Jacques est un délice qui se cuisine facilement. J'ai souhaité partager avec vous ces 30 recettes, mais il est évident qu'elle est source de bien d'autres créations. Un conseil : la noix de Saint-Jacques est bonne joueuse, à partir du moment où vous ne brutalisez pas sa chair délicate. Veillez à ne jamais trop la cuire : 2 minutes à la poêle suffisent amplement.

À vous de jouer ou d'improviser à partir de ces quelques idées, qui sauront, j'espère, vous régaler et vous inspirer.

Philippe Toinard

6

Tartare de bar et Saint-Jacques à la mangue

Préparation
25 minutes
+ 1 heure de marinade

Sans cuisson

Ingrédients pour 8 personnes
10 noix de Saint-Jacques
200 g de filets de bar
1 grosse mangue
1 échalote
2 cuillerées à soupe de ciboulette ciselée
3 cl de jus de yuzu
3 cuillerées à soupe d'huile d'olive
Sel, poivre du moulin

/ Taillez au couteau les filets de bar et les noix de Saint-Jacques en petits dés. Déposez le tout dans un saladier.

/ Pelez et émincez finement l'échalote. Mettez-la dans le saladier. Ajoutez la ciboulette et mélangez.

/ Coupez la mangue en deux et retirez le noyau. Incisez profondément la chair en forme de quadrillage. Appuyez sur la partie bombée de la mangue : les cubes vont se hérisser. Récupérez-les et taillez-les en petits dés. Ajoutez-les dans le saladier. Salez et poivrez.

/ Dans un bol, préparez une vinaigrette avec l'huile d'olive et le jus de yuzu. Versez-la dans le saladier. Mélangez et laissez mariner 1 heure au réfrigérateur avant de servir dans des petits ramequins.

Bon à savoir !
Vous pouvez remplacer le jus de yuzu par celui de 1 citron jaune.

8

Marinade de Saint-Jacques et granny smith

Préparation
25 minutes
+ 15 minutes de marinade

Sans cuisson

Ingrédients pour 8 personnes
8 noix de Saint-Jacques
3 pommes granny smith
2 endives
Le jus de 2 citrons verts
2 cuillerées à soupe d'huile d'olive
Fleur de sel
Poivre du moulin

/ Taillez au couteau les noix de Saint-Jacques en petits dés. Mettez-les dans un saladier.

/ Épluchez et épépinez les pommes puis taillez-les en dés d'une taille équivalente aux dés de Saint-Jacques. Ajoutez-les dans le saladier et arrosez le tout avec le jus des citrons. Laissez mariner 15 minutes.

/ Ajoutez l'huile d'olive dans le saladier. Assaisonnez de fleur de sel et de poivre. Mélangez et réservez au réfrigérateur.

/ Peu avant de servir, effeuillez les endives et conservez les pétales les plus incurvés. Rincez-les et essuyez-les bien avec un papier absorbant.

/ À l'aide d'une petite cuillère, répartissez la marinade de Saint-Jacques aux pommes dans chaque pétale d'endive et servez aussitôt.

Bon à savoir !
Vous pouvez parsemer la préparation d'une pincée de poudre de noix.

10

Verrines de Saint-Jacques, crabe et pomélo

Préparation
20 minutes

Sans cuisson

Ingrédients pour 4 verrines
8 noix de Saint-Jacques
150 g de chair de crabe
1 pomélo

Pour la mayonnaise
1 jaune d'œuf
1 cuillerée à café de moutarde
25 cl d'huile neutre (colza ou tournesol)
1 filet de vinaigre
Sel, poivre du moulin

/ Préparez la mayonnaise : dans un bol, mélangez le jaune d'œuf et la moutarde. Ajoutez 1 pincée de sel et de poivre. Versez l'huile en très mince filet en fouettant modérément jusqu'à ce que la mayonnaise devienne crémeuse et bien épaisse. Ajoutez un filet de vinaigre et rectifiez si besoin l'assaisonnement.

/ Taillez au couteau les noix de Saint-Jacques en petits dés. Mettez-les dans un saladier.

/ Coupez le pomélo en deux et détachez les quartiers. Mettez-les dans le saladier. Ajoutez la chair de crabe émiettée puis la mayonnaise. Mélangez et réservez au réfrigérateur.

/ Quinze minutes avant de servir, sortez le saladier et laissez-le à température ambiante.

/ Répartissez dans des verrines et servez.

Bon à savoir !
Pour réussir votre mayonnaise, le jaune d'œuf, la moutarde et l'huile doivent être à température ambiante (plus les ingrédients seront froids, plus la mayonnaise aura du mal à monter).

12

Cuillères de yaourt à l'ananas et dés de Saint-Jacques

Préparation
20 minutes

Cuisson
1 minute

Ingrédients pour 8 cuillères apéritives
8 noix de Saint-Jacques
½ ananas frais
2 yaourts nature
Piment d'Espelette
1 noix de beurre
Fleur de sel

/ Épluchez l'ananas et enlevez les yeux à l'aide d'un petit couteau pointu.
/ Coupez l'ananas en tranches. Ôtez le cœur et taillez la chair en petits dés.
/ Versez les yaourts dans un saladier. Ajoutez les dés d'ananas et mélangez. Réservez au frais.
/ Taillez au couteau les noix de Saint-Jacques en petits dés.
/ Faites fondre le beurre dans une poêle. Quand il est bien chaud, mettez les dés de Saint-Jacques à cuire 1 minute à peine à feu vif en remuant.
/ Dans les cuillères apéritives, versez le mélange yaourt-ananas. Répartissez dessus quelques dés de Saint-Jacques. Ajoutez une minuscule pincée de piment d'Espelette et 1 ou 2 cristaux de fleur de sel. Servez aussitôt.

Bon à savoir !
Si vous voulez que l'ensemble soit un peu plus relevé, ajoutez un peu de poivre dans le mélange ananas-yaourt.

14

Beignets de tempura de Saint-Jacques, sauce tartare

Préparation
20 minutes

Cuisson
2 à 3 minutes

Ingrédients pour 8 beignets
8 noix de Saint-Jacques
Huile de friture

Pour la pâte à beignets
65 g de farine
40 g de fécule de pomme de terre
7 g de levure chimique
5 g d'huile d'olive

Pour la sauce tartare
1 bol de mayonnaise maison
Une dizaine de cornichons
Quelques câpres
1 bouquet de persil plat
3 cuillerées à soupe de vinaigre
Sel, poivre du moulin

/ Préparez la pâte à beignets : tamisez la farine, la levure et la fécule dans un saladier. Ajoutez l'huile d'olive puis versez 125 g d'eau glacée en intégrant bien tous les ingrédients de façon à obtenir une pâte souple et sans grumeaux.

/ Pour la sauce tartare, mélangez à votre mayonnaise maison les cornichons taillés en petits dés, les câpres coupées en deux et le persil grossièrement haché. Déliez avec le vinaigre. Salez et poivrez. Réservez au frais.

/ Dans un wok, faites chauffer l'huile de friture à 180 °C.

/ Enfilez chaque noix de Saint-Jacques sur une pique à brochette en bois puis trempez-la dans la pâte à beignets.

/ Plongez les brochettes une par une dans la friture en évitant qu'elles ne s'agglutinent les unes aux autres. Lorsque les beignets sont bien gonflés et dorés (2 à 3 minutes), retirez-les avec une écumoire et égouttez-les sur du papier absorbant.

/ Servez les beignets chauds avec la sauce tartare.

Bon à savoir !
Pour vous assurer que l'huile est à bonne température, laissez tomber dans l'huile 1 goutte de pâte à beignets : si celle-ci crépite légèrement, c'est que l'huile est suffisamment chaude.

16

Rouleaux façon maki aux Saint-Jacques et saumon

Préparation
10 minutes

Cuisson
10 minutes

Ingrédients pour 4 rouleaux
4 feuilles de riz de 28 cm de diamètre
100 g de noix de Saint-Jacques
100 g de saumon cru
100 g de riz parfumé thaï
4 feuilles de salade
1 bouquet de coriandre
1 botte de basilic thaï
1 sachet de graines de sésame au wasabi
Vinaigre japonais
Sauce soja

/ Faites cuire le riz dans une casserole d'eau bouillante avec 2 cuillerées à soupe de vinaigre japonais pendant 10 minutes. Égouttez-le.
/ Lavez la salade et les herbes. Taillez au couteau les noix de Saint-Jacques en petits dés. Coupez le saumon en 4 lanières.
/ Passez 1 feuille de riz sous l'eau tiède pour la ramollir et placez-la sur le plan de travail. Saupoudrez-la de graines de sésame au wasabi.
/ Sur le premier tiers de la feuille, disposez le riz en rectangle. Ajoutez par-dessus quelques feuilles de basilic et de coriandre.
/ Derrière le riz posez 1 feuille de salade (entière ou grossièrement hachée selon sa taille).
/ Derrière la salade, placez 1 lanière de saumon et quelques dés de Saint-Jacques.
/ Roulez la feuille de riz en appuyant fermement. Lorsque le saumon et les dés de Saint-Jacques sont intégrés dans le rouleau, rabattez au centre chaque côté de la feuille en formant un angle droit puis finissez de rouler.
/ Réalisez ainsi les autres rouleaux.
/ Coupez chaque rouleau en 6 morceaux. Dégustez aussitôt avec la sauce soja (diluée avec un peu d'eau si vous la trouvez trop salée).

Bon à savoir !
Vous pouvez remplacer le saumon par du thon.

18

Saint-Jacques snackées au bouillon d'herbes

Préparation
15 minutes

Cuisson
5 minutes

Ingrédients pour 4 personnes
16 noix de Saint-Jacques
1 cube de bouillon de légumes
1 botte de ciboulette
½ bouquet de coriandre
½ bouquet de cerfeuil
2 cuillerées à soupe d'huile d'olive
Fleur de sel
Poivre du moulin

/ Dans une casserole, portez à ébullition 1 litre d'eau avec le cube de bouillon de légumes.
/ Lavez et séchez les herbes avant de les hacher grossièrement au couteau.
/ Dans une poêle, faites chauffer l'huile d'olive puis faites dorer les Saint-Jacques 1 à 2 minutes sur chaque face. Déposez-les sur du papier absorbant pour éliminer l'excès de graisse.
/ Poivrez, salez et recouvrez les Saint-Jacques d'une assiette creuse pour qu'elles restent chaudes.
/ Incorporez les herbes hachées dans le bouillon. Attendez une nouvelle ébullition et retirez la casserole du feu.
/ Dans des assiettes creuses, disposez 4 noix de Saint-Jacques. Recouvrez de bouillon aux herbes et dégustez aussitôt.

Bon à savoir !
Vous pouvez ajouter dans le bouillon 5 à 6 pistils de safran.

20

Crème de potiron et Saint-Jacques snackées

Préparation
20 minutes

Cuisson
35 minutes

Ingrédients pour 4 personnes
16 noix de Saint-Jacques
1 potiron de petite taille
1 cube de bouillon de volaille
2 oignons
2 gousses d'ail
1 bouquet garni
3 cuillerées à soupe de crème fraîche
Huile de noisette
4 cuillerées à soupe d'huile d'olive
Sel, poivre du moulin

/ Dans une casserole, portez à ébullition 50 cl d'eau avec le cube de bouillon de volaille.

/ Épluchez le potiron, enlevez les graines et les filaments, puis détaillez la chair en gros cubes.

/ Pelez et hachez les oignons et les gousses d'ail. Faites suer l'ensemble dans une sauteuse avec 2 cuillerées à soupe d'huile d'olive. Ajoutez les cubes de potiron et le bouquet garni. Mouillez à hauteur avec le bouillon. Couvrez et laissez cuire 30 minutes à feu doux.

/ Retirez le bouquet garni. Salez, poivrez et ajoutez la crème fraîche. Mélangez puis mixez finement. Remettez la crème de potiron quelques minutes sur feu doux.

/ Dans une poêle, faites chauffer le reste d'huile d'olive puis faites dorer les Saint-Jacques 1 à 2 minutes sur chaque face. Déposez-les sur du papier absorbant pour éliminer l'excès de graisse.

/ Versez la crème de potiron dans des assiettes creuses. Ajoutez 4 noix de Saint-Jacques. Assaisonnez de poivre et d'un filet d'huile de noisette.

Bon à savoir !
Si ce n'est pas la saison du potiron, présentez vos Saint-Jacques sur une crème aux champignons de Paris.

22

Carpaccio de Saint-Jacques et tartare de lotte

Préparation
25 minutes
+ 1 heure de marinade

Sans cuisson

Ingrédients pour 4 personnes
8 noix de Saint-Jacques
400 g de lotte
1 oignon rouge
1 bouquet de persil plat
1 bouquet de coriandre
Le jus de 3 citrons verts
Le jus de 1 citron jaune
2 cuillerées à soupe d'huile d'olive
Sel, poivre du moulin

/ Parez la lotte pour ne conserver que la chair blanche. Coupez-la en petits dés.
/ Lavez et ciselez finement la coriandre et le persil. Épluchez et hachez l'oignon.
/ Dans un saladier, mélangez délicatement les dés de lotte avec la coriandre, le persil et l'oignon. Arrosez avec le jus des citrons verts. Salez, poivrez et laissez mariner 1 heure environ.
/ Disposez les noix de Saint-Jacques face contre face de façon à obtenir un boudin. Emballez ce boudin dans du film alimentaire bien serré et placez-le au congélateur pendant 30 minutes.
/ Dans un bol, mélangez le jus du citron jaune avec l'huile d'olive et 1 pincée de sel.
/ Sortez les Saint-Jacques du congélateur, enlevez le film alimentaire et découpez-les en fines lamelles avec un couteau bien aiguisé.
/ Dans chaque assiette disposez des lamelles de Saint-Jacques et 2 quenelles de tartare de lotte. Arrosez les Saint-Jacques de sauce au citron et donnez un tour de moulin à poivre.

Bon à savoir !
Vous pouvez remplacer la lotte par du bar ou du cabillaud.

24

Salade tiède de Saint-Jacques, couteaux et bulots

Préparation
40 minutes

Cuisson
10 minutes

Ingrédients pour 4 personnes
12 noix de Saint-Jacques
500 g de bulots
20 couteaux
400 g de petites pommes de terre type grenaille
1 oignon
3 cuillerées à soupe de crème liquide
1 bouquet garni
1 clou de girofle
1 cuillerée à café de moutarde
2 cuillerées à soupe de vinaigre
2 cuillerées à soupe d'huile d'olive
Sel
Poivre moulu

/ Portez à ébullition une grande casserole d'eau avec 4 belles pincées de poivre moulu. Ajoutez le bouquet garni, l'oignon piqué du clou de girofle et les bulots. Laissez cuire 7 à 10 minutes selon la taille des coquillages.

/ Faites cuire les pommes de terre avec leur peau dans une casserole d'eau bouillante salée pendant 10 minutes environ. Épluchez-les puis coupez-les en deux ou en trois.

/ Décortiquez les bulots et ôtez l'opercule noir. Rincez-les à l'eau tiède et coupez-les en dés.

/ Faites cuire les couteaux à la vapeur quelques secondes juste pour les ouvrir. Nettoyez-les avant de les éponger puis taillez-les en fine julienne.

/ Dans un bol, mélangez le vinaigre, la moutarde, la crème liquide, le sel et le poivre.

/ Dans une poêle, faites chauffer l'huile d'olive puis faites dorer les Saint-Jacques 1 à 2 minutes sur chaque face. Déposez-les sur du papier absorbant pour éliminer l'excès de graisse. Poivrez et coupez chaque noix en 8 petits morceaux.

/ Dans un saladier, mélangez délicatement les bulots, les pommes de terre, les couteaux et les Saint-Jacques.

/ Répartissez la salade de la mer dans de jolies assiettes. Présentez la vinaigrette à part.

Bon à savoir !
Pour vous faciliter la vie, vous pouvez acheter des bulots déjà cuits chez votre poissonnier.

Verrines de boudin noir, pommes et Saint-Jacques

Préparation
30 minutes

Cuisson
10 minutes

Ingrédients pour 4 verrines
8 noix de Saint-Jacques
400 g de pommes type golden
250 g de boudin noir
1 oignon
30 g de beurre
Sel, poivre du moulin

/ Disposez les noix de Saint-Jacques face contre face de façon à obtenir un boudin. Emballez ce boudin dans du film alimentaire bien serré et placez-le au congélateur pendant 30 minutes.
/ Épluchez les pommes, coupez-les en quartiers puis en cubes.
/ Dans une casserole, faites fondre la moitié du beurre avec 1 cuillerée à soupe d'eau puis déposez les cubes de pomme. Laissez cuire 10 minutes environ à feu doux.
/ Épluchez et hachez l'oignon. Enlevez la peau du boudin.
/ Dans une poêle, faites revenir l'oignon haché avec le reste de beurre. Ajoutez la chair de boudin et laissez cuire 5 à 7 minutes à feu doux. Salez et poivrez si vous le souhaitez.
/ Sortez les Saint-Jacques du congélateur, enlevez le film alimentaire et découpez-les en fines lamelles avec un couteau bien aiguisé.
/ Écrasez grossièrement les pommes à la fourchette. Dans chaque verrine, déposez une couche de boudin puis une couche de pommes. Recommencez l'opération puis terminez par des lamelles de Saint-Jacques disposées en rosace.

Bon à savoir !
Vous pouvez présenter cette recette dans l'assiette en vous servant d'un emporte-pièce et l'accompagner de mesclun.

28

Salade de Saint-Jacques poêlées, gouda vieux et mesclun

Préparation
10 minutes

Cuisson
2 minutes

Ingrédients pour 4 personnes
12 noix de Saint-Jacques
100 g de mesclun
(avec un peu de roquette)
120 g de gouda vieux
1 poignée de graines de pavot
1 poignée de graines de lin
1 poignée de graines de sésame
4 cuillerées à soupe d'huile d'olive
1 cuillerée à soupe de vinaigre de cidre
Fleur de sel
Sel, poivre du moulin

/ Lavez et essorez le mesclun.
/ Dans un saladier, mélangez 3 cuillerées à soupe d'huile d'olive et le vinaigre de cidre. Salez, poivrez et ajoutez les graines.
/ Râpez grossièrement le gouda à la mandoline.
/ Dans une poêle, faites chauffer le reste d'huile d'olive puis faites dorer les Saint-Jacques 1 à 2 minutes sur chaque face. Déposez-les sur du papier absorbant pour éliminer l'excès de graisse.
/ Dans de jolis bols, faites un nid de mesclun. Déposez les noix de Saint-Jacques au milieu, assaisonnez-les de fleur de sel et de poivre du moulin. Parsemez de gouda émietté.
/ Arrosez le tout de vinaigrette et servez aussitôt.

Bon à savoir !
L'association noix de Saint-Jacques et parmesan est aussi une alternative très alléchante.

Tartines de Saint-Jacques marinées

Préparation
40 minutes
+ 30 minutes de marinade

Sans cuisson

Ingrédients pour 4 tartines
4 belles tranches de pain de campagne
8 noix de Saint-Jacques
2 bulbes de fenouil
2 cuillerées à soupe d'aneth ciselé
1 cuillerée à soupe de basilic ciselé
10 baies de genièvre
Le jus de 1 citron vert
Le jus de 1 citron jaune
5 cl de vodka
1 cuillerée à soupe de moutarde
2 cuillerées à soupe d'huile d'olive
1 cuillerée à café de vinaigre
1 cuillerée à soupe de sucre semoule
Sel, poivre du moulin

/ Disposez les noix de Saint-Jacques face contre face de façon à obtenir un boudin. Emballez ce boudin dans du film alimentaire bien serré et placez-le au congélateur pendant 30 minutes.

/ Dans un saladier, mélangez la moutarde, le jus du citron jaune, le vinaigre et l'huile d'olive. Salez et poivrez.

/ Supprimez les petites feuilles du fenouil ainsi que les extrémités vertes. Rincez les bulbes. Coupez-les en deux, enlevez le trognon et taillez-les en lanières. Mélangez-les à la vinaigrette dans le saladier.

/ Préparez la marinade : hachez les baies de genièvre et mettez-les dans un bol. Ajoutez l'aneth, le basilic, le sucre, le jus du citron vert et la vodka.

/ Sortez les Saint-Jacques du congélateur, ôtez le film alimentaire et découpez-les en fines lamelles avec un couteau bien aiguisé. Disposez-les dans un plat creux et arrosez-les avec la marinade aux herbes. Laissez reposer 30 minutes.

/ Enlevez la croûte des tranches de pain de façon bien nette et toastez les rectangles obtenus.

/ Égouttez les Saint-Jacques de la marinade et disposez-les sur les tranches de pain tièdes. Ajoutez 1 cuillerée à soupe de lanières de fenouil.

Bon à savoir !
La marinade peut être préparée la veille et conservée au réfrigérateur.

Carpaccio de Saint-Jacques, chutney de fruits exotiques

Préparation
30 minutes
+ 30 minutes de marinade

Cuisson
1 heure (pour le chutney)

Ingrédients pour 4 personnes
16 noix de Saint-Jacques
1 papaye
1 mangue
2 fruits de la passion
2 oignons rouges
1 cuillerée à soupe de basilic ciselé
Le jus de 1 citron vert
12 cuillerées à soupe d'huile d'olive
2 cuillerées de vinaigre balsamique blanc
2 belles cuillerées à soupe de miel
Sel, poivre du moulin

/ Pelez les oignons et émincez-les finement. Épluchez et coupez en dés la papaye et la mangue. Récupérez la chair et les graines des fruits de la passion.

/ Dans une sauteuse, faites suer les oignons avec 2 cuillerées à soupe d'huile d'olive 5 minutes à feu doux. Incorporez le miel et laissez cuire 2 minutes. Ajoutez tous les fruits et le vinaigre balsamique. Couvrez et laissez compoter 1 heure à feu doux.

/ Disposez les noix de Saint-Jacques face contre face de façon à obtenir un boudin. Emballez ce boudin dans du film alimentaire bien serré et placez-le au congélateur pendant 30 minutes.

/ Dans un bol, mélangez le reste d'huile d'olive, le jus de citron vert et le basilic. Salez et poivrez.

/ Sortez les Saint-Jacques du congélateur, enlevez le film alimentaire et découpez-les en fines lamelles. Disposez-les dans un plat creux et arrosez-les de sauce au citron vert. Laissez mariner 30 minutes.

/ Égouttez les Saint-Jacques de la marinade et disposez-les harmonieusement dans chaque assiette. Ajoutez 1 quenelle de chutney encore tiède et dégustez aussitôt.

Bon à savoir !
Pour varier, vous pouvez réaliser un chutney avec de l'ananas, des physalis, des oranges et des pamplemousses.

34 Tartes fines aux cèpes et Saint-Jacques

Préparation
30 minutes
+ 30 minutes de congélation

Cuisson
15 minutes

Ingrédients pour 4 personnes
1 rouleau de pâte feuilletée
12 noix de Saint-Jacques
6 cèpes de taille moyenne
1 jaune d'œuf
1 oignon
1 gousse d'ail
1 bouquet de persil plat
1 cuillerée à soupe de noisettes pilées
50 g de beurre
10 cl d'huile d'olive
Huile de truffe
Fleur de sel
Poivre du moulin

/ Disposez les noix de Saint-Jacques face contre face de façon à obtenir un boudin. Emballez ce boudin dans du film alimentaire bien serré et placez-le au congélateur pendant 30 minutes.

/ Préchauffez le four à 180 °C (th. 6).

/ Lavez et hachez le persil. Nettoyez les cèpes à l'aide d'un pinceau et ôtez le bout terreux. Tranchez les têtes en fines lamelles et réservez. Coupez les pieds en petits dés.

/ Épluchez et hachez grossièrement l'ail et l'oignon. Faites-les suer ensemble dans une poêle à feu doux avec le beurre et 1 cuillerée à soupe d'huile d'olive jusqu'à ce qu'ils deviennent translucides. Ajoutez les dés de cèpes et le persil haché. Laissez cuire 5 minutes.

/ Déroulez la pâte feuilletée et coupez 4 carrés de 15 cm de côté. Placez-les sur une plaque à pâtisserie recouverte de papier sulfurisé et badigeonnez-les de jaune d'œuf délayé avec un peu d'eau.

/ Au centre de chaque carré, déposez un peu du mélange cèpes-persil-oignons. Ajoutez un filet d'huile d'olive et quelques cristaux de fleur de sel. Donnez un tour de moulin à poivre. Enfournez et laissez cuire 15 minutes.

/ Sortez les Saint-Jacques du congélateur, enlevez le film alimentaire et découpez-les en fines lamelles.

/ Sur chaque tarte garnie, disposez, en les intercalant, des lamelles de cèpes crues et de Saint-Jacques. Juste avant de servir, arrosez d'un filet d'huile de truffe et parsemez de noisettes pilées.

Saint-Jacques poêlées et risotto au parmesan

Préparation
20 minutes

Cuisson
20 minutes

Ingrédients pour 4 personnes
20 petites noix de Saint-Jacques
4 belles poignées de riz type arborio
1 oignon
100 g de parmesan râpé
1 cube de bouillon de légumes
½ verre de vin blanc sec
100 g de beurre ramolli
4 cuillerées à soupe d'huile d'olive
Fleur de sel
Poivre du moulin

/ Portez à ébullition 1 litre d'eau dans une casserole et diluez le cube de bouillon.
/ Épluchez et émincez l'oignon.
/ Faites chauffer un wok (ou une sauteuse) avec 2 cuillerées à soupe d'huile d'olive, puis faites légèrement colorer l'oignon à feu doux. Ajoutez le riz et laissez-le cuire 3 à 4 minutes en remuant jusqu'à ce qu'il devienne translucide.
/ Versez le vin blanc dans le wok. Laissez-le s'évaporer puis ajoutez le bouillon, louche après louche, en mélangeant bien (attendez qu'il soit totalement absorbé avant d'en rajouter). Faites ainsi cuire le riz 20 minutes environ à feu doux sans jamais cesser de remuer.
/ Poivrez généreusement et incorporez délicatement le beurre ramolli puis le parmesan râpé. Réservez au chaud.
/ Dans une poêle, faites chauffer le reste d'huile d'olive puis faites légèrement dorer les Saint-Jacques 1 à 2 minutes sur chaque face. Déposez-les sur du papier absorbant pour éliminer l'excès de graisse.
/ Disposez 5 noix de Saint-Jacques dans chaque assiette creuse. Assaisonnez de fleur de sel et de poivre du moulin. Ajoutez le risotto et servez aussitôt.

Bon à savoir !
Le bouillon et le parmesan étant salés, il n'est pas nécessaire d'ajouter du sel dans la préparation d'un risotto.

Saint-Jacques panées à la poudre de noisettes, purée de pommes de terre

Préparation
20 minutes

Cuisson
20 minutes

Ingrédients pour 4 personnes
12 noix de Saint-Jacques
100 g de poudre de noisettes
4 cuillerées à soupe d'huile d'olive
Sel, poivre du moulin

Pour la purée de pommes de terre
1 kg de pommes de terre type bintje
20 cl de lait
1 pincée de noix de muscade râpée
120 g de beurre
Gros sel
Sel, poivre du moulin

/ Épluchez les pommes de terre et coupez-les en deux ou en quatre selon la grosseur. Mettez-les dans une casserole avec une bonne pincée de gros sel. Mouillez d'eau à hauteur et faites cuire 20 minutes à feu moyen.
/ Passez les pommes de terre au presse-purée.
/ Versez la purée dans une casserole. Ajoutez le lait, la muscade râpée et 1 pincée de sel. Portez à feu doux et mélangez délicatement pendant 1 minute. Incorporez le beurre en dés et remuez jusqu'à ce qu'il soit totalement fondu. Poivrez généreusement. Réservez au chaud.
/ À l'aide d'un pinceau, badigeonnez les noix de Saint-Jacques d'huile d'olive. Salez et poivrez. Roulez-les dans la poudre de noisettes.
/ Dans une poêle, faites chauffer 2 cuillerées à soupe d'huile d'olive puis faites légèrement dorer les Saint-Jacques 1 à 2 minutes sur chaque face. Déposez-les sur du papier absorbant pour éliminer l'excès de graisse.
/ Dressez 2 quenelles de purée dans chaque assiette. Ajoutez 3 noix de Saint-Jacques et servez aussitôt.

Bon à savoir !
Vous pouvez aussi paner les noix de Saint-Jacques avec de la poudre d'amandes ou un mélange poudre de noisettes-poudre d'amandes.

40

Brochettes de Saint-Jacques à la citronnelle, vinaigrette fruit de la passion

Préparation
15 minutes

Cuisson
3 minutes

Ingrédients pour 4 personnes
16 noix de Saint-Jacques
4 belles poignées de mesclun
3 fruits de la passion
4 tiges de citronnelle
Le jus de 1 citron vert
10 cl de lait de coco
8 cuillerées à soupe d'huile d'olive
Sel de céleri
Sel, poivre du moulin

/ Coupez les fruits de la passion en deux. Récupérez le jus et les graines à l'aide d'une petite cuillère.

/ Dans un bol, mélangez au fouet 5 cuillerées à soupe d'huile d'olive, le jus de citron vert et le lait de coco. Ajoutez le jus et les graines des fruits de la passion. Salez et poivrez.

/ Taillez en biseau la pointe des tiges de citronnelle. Piquez 4 noix de Saint-Jacques sur chaque tige.

/ À l'aide d'un pinceau, badigeonnez légèrement les brochettes avec le reste d'huile d'olive puis faites-les dorer sur le gril 1 min 30 de chaque côté.

/ Disposez chaque brochette sur une assiette. Assaisonnez de sel de céleri puis arrosez d'1 cuillerée à soupe de vinaigrette aux fruits de la passion.

/ Servez chaque brochette avec 1 poignée de mesclun assaisonnée du reste de vinaigrette.

Bon à savoir !
Vous pouvez ajouter à votre mesclun un peu de coriandre ou de menthe hachée.

42 Millefeuilles de Saint-Jacques au chorizo et lard de Colonnata

Préparation
10 minutes

Cuisson
5 minutes

Ingrédients pour 4 personnes
16 noix de Saint-Jacques
20 g de lard de Colonnata coupé en 16 fines tranches
32 rondelles de chorizo
3 cuillerées à soupe d'huile d'olive
Fleur de sel
Poivre
4 belles poignées de mesclun pour servir

/ Poêlez 16 rondelles de chorizo 30 secondes à peine de chaque côté dans un filet d'huile d'olive. Débarrassez-les sur une assiette et recouvrez d'une seconde assiette pour les maintenir au chaud.

/ Faites chauffer 2 cuillerées à soupe d'huile d'olive dans une autre poêle, puis faites légèrement dorer les noix de Saint-Jacques 1 à 2 minutes de chaque côté à feu doux. Déposez-les sur du papier absorbant pour éliminer l'excès de graisse. Poivrez, salez, puis réservez-les au chaud avec les rondelles de chorizo frites.

/ Poêlez les rondelles de chorizo restantes 30 secondes à peine de chaque côté dans un filet d'huile d'olive puis réservez-les avec les premières.

/ Coupez les Saint-Jacques à mi-hauteur.

/ Montez les millefeuilles : dans chaque assiette, disposez 1 rondelle de chorizo pour former le socle. Ajoutez ½ noix de Saint-Jacques, 1 rondelle de chorizo puis la seconde moitié de Saint-Jacques. Placez sur le dessus 1 tranche de lard et laissez fondre au contact de la chaleur. Dressez ainsi 4 millefeuilles par assiette.

/ Servez aussitôt avec 1 poignée de mesclun assaisonnée de vinaigrette.

44 Saint-Jacques sur lit de poireaux, jus aux agrumes

Préparation
30 minutes

Cuisson
25 minutes

Ingrédients pour 4 personnes
16 noix de Saint-Jacques
(avec 8 coquilles vides)
2 blancs de poireaux
2 cuillerées à soupe de crème fraîche
2 échalotes
1 bouquet de persil plat
1 pincée de noix de muscade râpée
Le jus de 6 oranges
Le jus de 1 citron
10 cl de vin blanc sec
50 g de beurre
Fleur de sel
Sel, poivre du moulin

/ Nettoyez les blancs de poireaux et émincez-les finement.

/ Dans une sauteuse, faites fondre 30 g de beurre puis faites suer les poireaux à petit feu jusqu'à ce qu'ils deviennent translucides. Salez, poivrez, ajoutez le vin blanc et la crème fraîche. Mélangez et laissez réduire 5 minutes environ à feu doux en remuant de temps en temps. Réservez au chaud.

/ Pelez et émincez les échalotes. Rincez et hachez le persil.

/ Versez le jus des agrumes dans une casserole. Ajoutez les échalotes. Assaisonnez de sel, de poivre et de muscade râpée. Laissez réduire 20 minutes à feu doux.

/ Dans une poêle, faites fondre le reste de beurre, puis faites légèrement dorer les Saint-Jacques 1 à 2 minutes sur chaque face à feu doux. Déposez-les sur du papier absorbant pour éliminer l'excès de graisse.

/ Dans chaque coquille, déposez un peu de fondue de poireaux et placez 2 noix de Saint-Jacques par-dessus. Poivrez et ajoutez quelques cristaux de fleur de sel. Arrosez de jus aux agrumes et décorez de persil. Servez aussitôt.

Bon à savoir !
Au lieu de cuire les Saint-Jacques à la poêle, vous pouvez aussi les disposer sur la fondue de poireaux et passer les coquilles 2 minutes sous le gril du four ; arrosez alors de jus aux agrumes juste avant de servir.

Crumbles de Saint-Jacques, roquette au parmesan

Préparation
20 minutes

Cuisson
10 minutes

Ingrédients pour 4 personnes
16 noix de Saint-Jacques
300 g de roquette
50 g de parmesan
100 g de mie de pain
1 belle poignée de pignons de pin
50 g de poudre d'amandes
50 g de poudre de noisettes
50 g de beurre pommade
2 cuillerées à soupe d'huile d'olive
Sel, poivre du moulin

Pour la vinaigrette
3 cuillerées à soupe d'huile d'olive
2 cuillerées à soupe de vinaigre
Sel, poivre du moulin

/ Préchauffez le four à 240 °C (th. 8).

/ Préparez la pâte à crumble : dans un saladier, mélangez les poudres d'amandes et de noisettes, la mie de pain, 2 pincées de sel et le beurre pommade. Travaillez la pâte jusqu'à ce qu'elle devienne sableuse et s'émiette facilement.

/ Dans une poêle, faites chauffer l'huile d'olive puis faites légèrement dorer les noix de Saint-Jacques 1 à 2 minutes de chaque côté. Déposez-les sur du papier absorbant pour éliminer l'excès de graisse. Donnez un tour de moulin à poivre sur chaque noix.

/ Coupez les noix en quatre morceaux et disposez-les dans chaque cassolette. Saupoudrez de pâte à crumble.

/ Enfournez et laissez cuire 10 minutes.

/ Dans un bol, mélangez tous les ingrédients de la vinaigrette.

/ Lavez la roquette et mettez-la dans un saladier. Ajoutez la vinaigrette, les pignons et des lamelles de parmesan prélevées à l'aide d'un économe. Mélangez le tout. Servez avec les crumbles.

Saint-Jacques, crème de panais et poitrine séchée

Préparation
25 minutes

Cuisson
25 minutes

Ingrédients pour 4 personnes
16 noix de Saint-Jacques
600 g de panais
4 tranches de poitrine fumée
1 cube de bouillon de volaille
10 cl de crème liquide
100 g de beurre
2 cuillerées à soupe d'huile d'olive
Sel, poivre du moulin

/ Préchauffez le four à 240 °C (th. 8).

/ Épluchez et lavez les panais ; coupez-les en gros morceaux. Déposez-les dans une casserole et couvrez d'eau à hauteur. Salez et faites cuire 25 minutes à feu doux.

/ Diluez le cube de bouillon de volaille dans 50 cl d'eau bouillante.

/ Déposez les tranches de poitrine sur une plaque. Enfournez et faites griller 5 minutes environ en surveillant bien la cuisson : elles doivent être presque cassantes.

/ Dans une poêle, faites chauffer l'huile d'olive, puis faites légèrement dorer les noix de Saint-Jacques 1 à 2 minutes de chaque côté. Déposez-les sur du papier absorbant pour éliminer l'excès de graisse. Donnez un tour de moulin à poivre sur chaque noix.

/ Égouttez les panais et mixez-les en ajoutant petit à petit le bouillon de volaille, la crème liquide et le beurre coupé en dés. Salez et poivrez.

/ Retirez les tranches de poitrine du four et émiettez-les grossièrement.

/ Dans chaque assiette creuse, disposez 4 noix de Saint-Jacques sur la crème de panais et parsemez de miettes de poitrine grillée.

Bon à savoir !
Vous pouvez accélérer la cuisson des tranches de poitrine fumée en mettant votre four sur position gril.
Les panais s'oxydent très vite : cuisez-les aussitôt après les avoir épluchés.

Saint-Jacques poêlées et compotée d'oignons au miel

Préparation
15 minutes

Cuisson
20 minutes

Ingrédients pour 4 personnes
16 noix de Saint-Jacques avec 8 coquilles vides
5 oignons
1 poignée de raisins secs
4 cuillerées à soupe d'huile d'olive
1 cuillerée à soupe de sucre semoule
3 cuillerées à soupe de miel liquide
Fleur de sel
Poivre du moulin
4 belles poignées de mesclun pour servir

Pour la vinaigrette
3 cuillerées à soupe d'huile d'olive
2 cuillerées à soupe de vinaigre
1 cuillerée à café de miel liquide
Sel, poivre du moulin

- Épluchez et émincez les oignons. Faites-les revenir à feu moyen dans une sauteuse avec 2 cuillerées à soupe d'huile d'olive jusqu'à ce qu'ils soient bien dorés. Ajoutez ½ verre d'eau, le sucre, le miel puis les raisins secs. Couvrez et laissez cuire 20 minutes environ à feu doux.
- Dans un saladier, mélangez tous les ingrédients de la vinaigrette.
- Lavez le mesclun et mélangez-le à la vinaigrette.
- Dans une poêle, faites chauffer le reste d'huile d'olive puis faites légèrement dorer les noix de Saint-Jacques 1 à 2 minutes de chaque côté. Déposez-les sur du papier absorbant pour éliminer l'excès de graisse. Ajoutez un tour de moulin à poivre et quelques cristaux de fleur de sel sur chaque noix.
- Dans le fond de chaque coquille, déposez un peu de compotée d'oignons puis ajoutez 2 noix de Saint-Jacques par-dessus. Servez le mesclun à part.

Bon à savoir !
Vous pouvez préparer davantage de compotée d'oignons et la conserver jusqu'au lendemain au réfrigérateur. Elle accompagne merveilleusement bien le foie gras, par exemple.

52

Ravioles de Saint-Jacques au thym, crème aux pommes

Préparation
30 minutes
+ 15 minutes au réfrigérateur

Cuisson
15 minutes

Ingrédients pour 4 personnes
12 noix de Saint-Jacques
2 feuilles de pâte à raviole
4 pommes type golden
1 jaune d'œuf
Quelques brindilles de thym
20 cl de crème liquide
40 g de beurre
1 filet d'huile d'olive
50 g de sucre semoule
Sel, poivre du moulin

/ Découpez les feuilles de pâte à raviole de manière à obtenir 24 carrés.
/ Au centre d'un carré, déposez 1 noix de Saint-Jacques. Salez, poivrez et parsemez de thym émietté.
/ À l'aide d'un pinceau, badigeonnez le pourtour de la raviole de jaune d'œuf délayé dans un peu d'eau. Couvrez la noix de Saint-Jacques d'un autre carré de pâte et soudez les deux carrés en appuyant sur les bords. Réalisez ainsi les autres ravioles et réservez 15 minutes au frais.
/ Lavez les pommes, coupez-les en quartiers et épépinez-les.
/ Dans une poêle, faites fondre le beurre puis faites revenir les pommes 5 minutes à feu doux. Saupoudrez de sucre et laissez caraméliser 5 minutes à feu vif. Ajoutez la crème liquide et mélangez sans briser les pommes.
/ Portez à frémissements une casserole d'eau avec l'huile d'olive.
/ À l'aide d'une écumoire, plongez les ravioles dans l'eau et laissez cuire 3 minutes. Égouttez-les délicatement.
/ Dans chaque assiette, préparez un nid de pommes à la crème et déposez 3 ravioles par-dessus. Dégustez aussitôt.

Bon à savoir !
Vous pouvez également cuire les ravioles 5 minutes au cuit-vapeur.

Ris de veau et Saint-Jacques, purée de pommes de terre

Préparation
20 minutes

Cuisson
25 minutes

Ingrédients pour 4 personnes
12 noix de Saint-Jacques
400 g de ris de veau parés
40 g de beurre
Fleur de sel
Sel, poivre du moulin

Pour la purée de pommes de terre
1 kg de pommes de terre type bintje
1 pincée de noix de muscade râpée
20 cl de lait
120 g de beurre
Gros sel
Sel, poivre du moulin

/ Préparez la purée : épluchez et lavez les pommes de terre ; coupez-les en quatre. Mettez-les dans une casserole avec une bonne pincée de gros sel. Mouillez d'eau à hauteur et faites cuire 20 minutes à feu moyen.

/ Égouttez les pommes de terre et passez-les au presse-purée.

/ Mettez la purée dans une casserole. Ajoutez le lait, la muscade râpée et 1 pincée de sel. Portez à feu doux et mélangez délicatement pendant 1 minute. Incorporez le beurre en dés et laissez-le fondre en remuant. Poivrez généreusement. Réservez au chaud.

/ Faites pocher les ris de veau 3 à 5 minutes dans une casserole d'eau bouillante. Passez-les sous l'eau froide puis tranchez-les en escalopes.

/ Dans une poêle, faites fondre 20 g de beurre puis faites revenir les ris de veau 1 minute de chaque côté à feu moyen. Salez et réservez au chaud.

/ Dans une autre poêle, faites fondre le reste de beurre puis faites légèrement dorer les noix de Saint-Jacques 1 à 2 minutes de chaque côté. Déposez-les sur du papier absorbant pour éliminer l'excès de graisse. Ajoutez 1 tour de moulin à poivre et quelques cristaux de fleur de sel sur chaque noix.

/ Dans chaque assiette dressez 3 noix de Saint-Jacques et 1 escalope de ris de veau. Accompagnez de purée.

Saint-Jacques lutées en coquilles, crème aux morilles

Préparation
20 minutes
+ 1 heure de trempage
pour les morilles

Cuisson
10 minutes

Ingrédients pour 4 personnes
12 noix de Saint-Jacques
(avec 4 coquilles vides :
les deux parties)
100 g de morilles déshydratées
1 rouleau de pâte feuilletée
1 échalote
½ bouquet de persil plat
25 cl de crème fleurette
10 cl de vin blanc sec
2 verres de lait
20 g de beurre
Sel, poivre du moulin

/ Faites tremper les morilles dans le lait pendant 1 heure environ pour les réhydrater.
/ Préchauffez le four à 240 °C (th. 8).
/ Épluchez et ciselez l'échalote. Hachez grossièrement le persil.
/ Égouttez les morilles.
/ Dans une casserole, faites fondre le beurre, puis ajoutez l'échalote ciselée et les morilles. Versez le vin et mélangez. Salez, poivrez et incorporez la crème fleurette. Laissez mijoter 5 minutes à feu doux. Ajoutez le persil en fin de cuisson.
/ Dans la pâte feuilletée, découpez 4 bandes de 2 cm de large environ.
/ Dans chaque coquille, déposez un peu de crème aux morilles puis placez par-dessus 3 noix de Saint-Jacques. Poivrez les noix.
/ Fermez les coquilles avec leur couvercle et lutez avec 1 bande de pâte feuilletée.
/ Enfournez et laissez cuire pendant 10 minutes.

Bon à savoir !
Vous pouvez réaliser cette recette avec toutes sortes de champignons déshydratés.

Papillotes de Saint-Jacques, petits légumes à la coriandre

Préparation
20 minutes

Cuisson
25 minutes

Ingrédients pour 4 personnes
16 noix de Saint-Jacques
4 carottes
2 blancs de poireaux
2 tomates
1 échalote
1 petit bouquet de coriandre
10 cl de crème liquide
10 cl de vin blanc
2 cuillerées à soupe d'huile d'olive
Sel, poivre du moulin

/ Préchauffez le four à 180 °C (th. 6).

/ Épluchez et taillez les carottes en gros dés. Émincez les blancs de poireaux. Coupez les tomates en quatre. Pelez et ciselez l'échalote.

/ Dans une poêle, faites chauffer l'huile d'olive, puis faites revenir l'échalote et les légumes 5 minutes à feu vif. Salez et poivrez. Couvrez et laissez mijoter 5 minutes à feu doux.

/ Découpez 4 carrés de papier cuisson de 20 cm de côté. Au centre de chaque carré, déposez un peu de préparation aux légumes mijotés puis placez 4 noix de Saint-Jacques par-dessus. Poivrez les noix et versez 1 filet de crème liquide. Ajoutez 2 à 3 feuilles de coriandre.

/ Fermez les papillotes en ménageant une ouverture ; versez-y un peu de vin blanc puis fermez hermétiquement.

/ Déposez les papillotes dans un plat allant au four huilé. Enfournez et laissez cuire 15 minutes.

Bon à savoir !
Les Saint-Jacques apprécient le parfum des herbes. À défaut de coriandre, vous pouvez parfumer vos papillotes avec du thym, du romarin ou du cerfeuil.

Chou farci aux Saint-Jacques et légumes

Préparation
25 minutes

Cuisson
25 minutes

Ingrédients pour 4 personnes
16 noix de Saint-Jacques
8 feuilles moyennes de chou frisé
150 g de champignons de Paris
2 carottes
2 blancs de poireaux
2 échalotes
3 cuillerées à soupe d'huile d'olive
Sel, poivre du moulin

/ Lavez les feuilles de chou puis blanchissez-les 10 minutes à l'eau bouillante. Égouttez-les et épongez-les sur du papier absorbant.
/ Préchauffez le four à 210 °C (th. 7).
/ Épluchez et taillez les carottes en gros dés. Enlevez le bout terreux des champignons, rincez-les rapidement sous l'eau froide et émincez-les.
/ Émincez les blancs de poireaux. Épluchez et ciselez les échalotes.
/ Dans une poêle, faites chauffer 2 cuillerées à soupe d'huile d'olive, puis faites revenir les échalotes et les légumes 5 minutes à feu vif. Couvrez et laissez mijoter 10 minutes à feu moyen.
/ Mixez les noix de Saint-Jacques et ajoutez-les au mélange. Salez et poivrez.
/ Au centre de chaque feuille de chou, déposez un peu de préparation aux légumes et Saint-Jacques. Rabattez les côtés de la feuille pour la fermer puis déposez-la dans un plat allant au four, la fermeture en dessous. Arrosez d'un filet d'huile d'olive.
/ Enfournez les feuilles de chou farcies et laissez cuire 10 minutes.

Bon à savoir !
Selon votre appétit, c'est la taille des feuilles qui détermine la quantité de farce : plus les feuilles sont grandes, plus il faudra préparer de farce.

Poêlée de Saint-Jacques, salade de betteraves, navets et pignons

Préparation
25 minutes

Cuisson
2 minutes

Ingrédients pour 4 personnes
16 noix de Saint-Jacques
2 navets
2 betteraves crues
1 salade verte
2 échalotes
1 belle poignée de pignons de pin
1 cuillerée à café de moutarde de Meaux
20 g de beurre
2 cuillerées à soupe d'huile d'olive
1 cuillerée à soupe de vinaigre de cidre
Fleur de sel
Sel, poivre du moulin

/ Dans un bol, mélangez la moutarde, l'huile d'olive et le vinaigre. Salez et poivrez. Ajoutez 1 échalote pelée et ciselée. Réservez.
/ Lavez la salade. Pelez et ciselez la seconde échalote.
/ Épluchez les betteraves et les navets. Râpez-les et mélangez-les dans un saladier. Ajoutez la salade et l'échalote. Assaisonnez de vinaigrette et mélangez le tout.
/ Dans une poêle, faites fondre le beurre, puis faites légèrement dorer les noix de Saint-Jacques 1 à 2 minutes de chaque côté. Déposez-les sur du papier absorbant pour éliminer l'excès de graisse. Donnez 1 tour de moulin à poivre et ajoutez quelques cristaux de fleur de sel sur chaque noix.
/ Répartissez la salade dans de jolis bols. Déposez 4 noix de Saint-Jacques par-dessus et parsemez de pignons de pin.

Bon à savoir !
Vous pouvez remplacer les navets crus par des radis type « green meat » ou « red meat », qui apporteront couleurs et saveurs.

Saint-Jacques gratinées

Préparation
25 minutes

Cuisson
20 minutes

Ingrédients pour 4 personnes
16 noix de Saint-Jacques
(avec 4 coquilles vides)
50 g de crevettes grises
100 g de champignons de Paris
1 échalote
2 cuillerées à soupe de farine
50 cl de crème fraîche
10 cl de vin blanc
40 g de beurre
Chapelure
Sel, poivre du moulin

/ Préchauffez le four à 180 °C (th. 6).

/ Décortiquez les crevettes. Pelez et ciselez l'échalote.

/ Enlevez le bout terreux des champignons. Rincez-les rapidement à l'eau fraîche puis émincez-les.

/ Faites fondre 20 g de beurre dans une sauteuse, puis faites revenir l'échalote 2 à 3 minutes à feu moyen. Ajoutez les crevettes et les champignons. Versez la farine et mélangez délicatement. Laissez cuire 2 minutes à feu moyen. Salez.

/ Dans une poêle, faites fondre le reste de beurre, puis faites légèrement dorer les noix de Saint-Jacques 1 à 2 minutes de chaque côté. Déposez-les sur du papier absorbant pour éliminer l'excès de graisse. Poivrez chaque noix.

/ Incorporez les noix de Saint-Jacques dans la sauteuse puis la crème fraîche et le vin blanc. Laissez mijoter 10 minutes à feu doux en remuant régulièrement. Rectifiez l'assaisonnement si nécessaire.

/ Répartissez la préparation dans les coquilles (4 noix par personne). Saupoudrez de chapelure.

/ Enfournez et laissez gratiner 5 à 7 minutes. Dégustez bien chaud.

Bon à savoir !
Si vous n'avez pas conservé les coquilles, faites cuire cette préparation dans un plat à gratin.

Remerciements

Je remercie Barbara Sabatier des éditions Mango pour son écoute et son enthousiasme, Adèle Vay pour son œil de lynx et évidemment toute l'équipe de maquettiste et correcteur.

Un grand merci à Isabel Brancq-Lepage pour l'interprétation de mes recettes qu'elle a malicieusement respectées. Enfin, merci à Fabrice Veigas, photographe de talent dont je ne connaissais pas le travail. Il me tarde de retravailler avec un tel duo.

Merci enfin à Floriane pour son soutien indéfectible, son palais aiguisé et son sens critique.